T0269160

CAMBRIDGE LIBRARY COLLECTION

Books of enduring scholarly value

Botany and Horticulture

Until the nineteenth century, the investigation of natural phenomena, plants and animals was considered either the preserve of elite scholars or a pastime for the leisured upper classes. As increasing academic rigour and systematisation was brought to the study of 'natural history', its subdisciplines were adopted into university curricula, and learned societies (such as the Royal Horticultural Society, founded in 1804) were established to support research in these areas. A related development was strong enthusiasm for exotic garden plants, which resulted in plant collecting expeditions to every corner of the globe, sometimes with tragic consequences. This series includes accounts of some of those expeditions, detailed reference works on the flora of different regions, and practical advice for amateur and professional gardeners.

Gleanings from French Gardens

The innovative gardener and writer William Robinson (1838–1935), several of whose other works are reissued in this series, was sent by The Times as its horticultural correspondent to the Paris International Exposition of 1867. As a result of his visit, he produced two books, The Parks, Promenades and Gardens of Paris (1869) and this highly illustrated work (first published in 1868 and reissued here in its 1869 second edition) on gardening trends in France, describing 'such features of French horticulture as are most worthy of adoption in British gardens'. In comparing French horticulture with British, Robinson believes that the gardens of the great houses of Britain are not matched in France, but that in terms of market gardening and its produce, France is definitely superior. He argues in this interesting work that French methods of training fruit such as apples, pears and peaches should be widely adopted.

Cambridge University Press has long been a pioneer in the reissuing of out-of-print titles from its own backlist, producing digital reprints of books that are still sought after by scholars and students but could not be reprinted economically using traditional technology. The Cambridge Library Collection extends this activity to a wider range of books which are still of importance to researchers and professionals, either for the source material they contain, or as landmarks in the history of their academic discipline.

Drawing from the world-renowned collections in the Cambridge University Library and other partner libraries, and guided by the advice of experts in each subject area, Cambridge University Press is using state-of-the-art scanning machines in its own Printing House to capture the content of each book selected for inclusion. The files are processed to give a consistently clear, crisp image, and the books finished to the high quality standard for which the Press is recognised around the world. The latest print-on-demand technology ensures that the books will remain available indefinitely, and that orders for single or multiple copies can quickly be supplied.

The Cambridge Library Collection brings back to life books of enduring scholarly value (including out-of-copyright works originally issued by other publishers) across a wide range of disciplines in the humanities and social sciences and in science and technology.

Gleanings from French Gardens

Comprising an Account of Such Features of French Horticulture as Are Most Worthy of Adoption in British Gardens

WILLIAM ROBINSON

CAMBRIDGE
UNIVERSITY PRESS

CAMBRIDGE
UNIVERSITY PRESS

University Printing House, Cambridge, CB2 8BS, United Kingdom

Cambridge University Press is part of the University of Cambridge.
It furthers the University's mission by disseminating knowledge in the pursuit of
education, learning and research at the highest international levels of excellence.

www.cambridge.org
Information on this title: www.cambridge.org/9781108079839

This edition first published 1869
This digitally printed version 2017

ISBN 978-1-108-07983-9 Paperback

MUSA ENSETE.

Aralia.   Phœnix.   Yucca.

GLEANINGS

FROM

FRENCH GARDENS:

COMPRISING AN

ACCOUNT OF SUCH FEATURES OF FRENCH HORTICULTURE AS
ARE MOST WORTHY OF ADOPTION IN BRITISH GARDENS.

By W. ROBINSON, F.L.S.

With Numerous Illustrations.

SECOND EDITION.

LONDON:
FREDERICK WARNE AND CO.
BEDFORD STREET, COVENT GARDEN.
NEW YORK: SCRIBNER, WELFORD, AND CO.
1869.

PREFACE.

SOME of the matters treated of in this book have lately
been the subjects of much discussion in the *Times* as well
as in all the horticultural papers; and to give the public a
fuller idea of them than could be gleaned from any or all of the
journals in which they were described or discussed, is my excuse for
writing a work which is so exceptional in its nature. I went
to France in January, 1867, with a view to study the horticulture
of the country so far as possible, while continuing my connexion
with the horticultural press; and in the course of the season I noticed
in the *Times*, the *Field*, and the *Gardener's Chronicle* such of the
features of French gardening as seemed to me worthy of adoption
in this country. In the correspondence which resulted from this,
each journal discussed a single phase of the subject or approached it
from its own point of view; and the want of illustrations prevented me from explaining in the most effective way several real
improvements in our gardens; thus it was difficult for the public
to get more than a vague notion of some of the matters of greatest
interest in French gardening. After the close of the discussion
in the horticultural journals, it occurred to me that a book, with
illustrations, would put the more important points in a clearer light;
and the result is the publication of the present volume. The work

is from its nature an imperfect one, and no less so for its object;
but I had no thought of writing it till some time after I had
quitted Paris, and in fact I was engaged in preparations for an
entirely different work, when the animated discussions on " French
and English gardening," and allied matters, obliged me to devote
myself almost entirely to the defence of what I am convinced are
the true and practical lessons of the matter.   No amount of interest excited on the part of the public would have induced me to
publish, were it not that I am certain we may adopt some of the
ways of our neighbours with decided advantage.

A few words as to the comparative merits of French and English
horticulture.   As regards the class of large private gardens, of which
we have such beautiful examples in many parts of this country,
there can be no comparison—there are few places in France which
equal ours.   In the culture of stove and greenhouse plants, of the
vine in glass-houses, of orchids, some vegetables, and in the general
keep and finish of their large gardens, they are quite behind us, not
to mention many minor matters.   The difference in the distribution of wealth in the two countries accounts for some of this in-
feriority; large gardens such as can only be supported by noblemen
and wealthy amateurs are far from being as plentiful as in this
country.   Their nurseries, and especially their plant nurseries, are
proportionately smaller; and among them you may look in vain
for such splendid establishments in that way as we possess about
London and in the provinces.   But when it comes to a supply
for their markets, and even for those of other countries, then I
am certain that they beat us; and I have never anywhere seen
such perfect examples of cultivation and rapid rotation as in the
Paris market gardens—not large, but with every span of the soil
at work, and green with abundant crops at all seasons.   In fruit
growing, they certainly lead; not always, as is commonly sup-

posed, from advantages of climate, but frequently under adverse circumstances. As for city-gardening, what has been done in Paris of late years, on the most magnificent scale ever attempted, is beyond all praise, and worthy of the best attention of all interested in town improvement. Finally, the graceful way the French decorate their apartments with plants, and develope beauty of vegetable form in connexion with brilliant flowers, is well worthy our imitation. However, as so many have during the past year seen their excellent public gardens, and particularly that at Passy, their well-supplied markets, and had a glimpse of their finished and careful fruit culture in the garden of the Exposition, I need hardly plead that we may learn something from the French. France is the orchard of Europe, and not only supplies us with enormous quantities of fruit, much of which we might ourselves grow in the southern counties of England and Ireland, but she also sends us quantities of salads in the winter and spring months, not grown in the mild and sunny regions, but abundantly in the colder parts, where the climate is quite as severe as our own; and if she only excelled in these two points they would be worthy of the attention of the British cultivator.

The climate requires a little consideration. Some have held that our "bad climate" would always prevent us from equalling the French as fruit-growers; others contended that it produced the finest fruits, the French being hard, gritty, sandy, &c.; while not a few have spoken of " the fine climate of France" as if it were some happy region in which the fruit-cultivator had merely to plant his trees and sit under them, nature doing the rest, and not a country the cliffs of whose coast are as plainly seen from our own shores as if they were those of one of our large estuaries. I am alluding to the climate of Paris and Northern France. France has several distinct climates; that of Paris does not differ so much from London as

London does from other parts of the British Isles, or even of England. Practically we are under almost the same influences; what one succeeds with the other may attempt with confidence. The Parisians get their finest peaches from within a few miles of the city, where they must be grown upon walls as in England and Ireland. Their finer winter pears they also grow upon sunny walls, and they take more pains to protect the blossoms in spring than we do. I well know that their climate is superior to ours for the culture of the pear, but am convinced that the difference in climate is not sufficient to account for our deficiency with this fine fruit; that with well managed and protected walls we can grow a supply of good winter pears for our own markets; and emphatically that as good and as cheap peaches may be grown in many parts of England and Ireland as ever were produced at Montreuil. Of one thing we may rest assured, that whatever the climate may do, the French fruit-grower does something more for his fruit trees than the British one generally does. The attention which is paid to equalizing the sap and keeping the tree thoroughly under control, and the success attained, are something quite remarkable. While I write this there is not a good fruit-garden in the neighbourhood of Paris but has its walls protected by a wide temporary coping, while in numerous cases with us there is no protection at all, or but a very imperfect one. This fact speaks for itself.

As regards the difference in the climate of the southern parts of England and the neighbourhood of Paris, considered with reference to what is called "subtropical gardening," I can offer some evidence. Last summer, I left Paris for six weeks from the month of August, solely with a view to examine the state of "subtropical gardening" in Britain and compare it with that around Paris. I came direct from Paris to the London parks, and was surprised to find very little difference, all· the more important fine-leaved plants looking in splendid health, one bed of Cannas at Battersea being nearly twelve feet high, the noble Wigandia in perfect vigour, and the Cannas, Ferdinandas, Polymnias, &c., even in the exposed Hyde Park and on the cold clay of the Regent's Park, quite healthy and vigorous. I even found some of them in presentable condition so far north as Archerfield, on the banks of the Firth of Forth. Considering how exceptionally bad was the past season, this proves that the growth of tender plants with large and effective foliage in the summer garden is by no means so impossible as is frequently supposed. Indeed the culture of the many hardy and half-hardy plants that may be used so advantageously to produce fine effects in the flower-garden, from elegant dwarf Conifers to graceful Bamboos, is much more easy in many parts of England than around Paris, where many subjects perish in winter that we find no difficulty in growing. The fact is, an infinitely more beautiful garden, even from the point of fine foliage, could be made in many mild spots near our coasts than is possible around Paris with any expense.

It now remains for me to thank my French friends for kind assistance rendered when in France and since leaving it, and among them M. H. Vilmorin, M. Jules Posth, M. E. André, MM. Newmann and Verlot, of the Jardin des Plantes, M. Ermens, of La Muette, M. Rose Charmeaux, of Thoméry, M. Jamain, of Bourg la Reine, M. Rivière, of the Luxembourg, M. Bergman, of Ferrières, Professor Du Breuil, M. Souchet, of Fontainebleau, and many others. Some of the cuts are from the "Atlas des Fleurs de Pleine Terre," of Vilmorin Andrieux and Co., some from "Culture des Arbres et Arbrisseux à Fruits de Table," of Professor Du Breuil, and a few are from the "Gardener's Chronicle." The rest are from designs furnished by the author.

As before indicated, this book does not pretend to deal fully with every feature of French horticulture, and where any reader, French or English, may notice the omission of an interesting or instructive point, the author will be much obliged by a communication on the subject being addressed to him, care of the Publishers, Bedford-street, Covent Garden, London, W.C.

W. ROBINSON.

LONDON, *April,* 1868.

CONTENTS.

———

LIST OF ILLUSTRATIONS.

List of Illustrations.

GLEANINGS FROM FRENCH GARDENS.

CHAPTER I.

Subtropical Gardening.

THE cultivation of plants distinguished by fine foliage or nobility of aspect, or in other words, the introduction of beauty of form to the flower garden, is a very desirable improvement that has been making its way in this country for a few years past. The system originated in Continental gardens, and has been brought to the greatest degree of perfection in and around Paris, where as well as in this country, the writer has had full opportunities of becoming acquainted with its merits. An attempt is accordingly here made to give an estimate of its value in relation to our own wants. The most desirable plants and families of plants which we can employ under this system are treated of rather fully, and concise selections given with descriptive notes, with a view of assisting every lover of a garden, no matter what his means, or where his situation in these islands, to add to his garden that which is so very rare in our " ornamental" ones—*beauty of form.* But it has been objected that the plants used for "subtropical gardening" are so tender and expensive that the system must be impossible to the greater number of British gardens. This is true of the sorts at present generally used, but why confine ourselves to these? No effect ever given in these latitudes by

B

costly "subtropicals" is superior to that which may be produced by
the use of hardy plants, combined with, where the climate admits,
such of the other class as are well known to be as cheaply raised,
and as free of growth as common " bedding stuff." Therefore, instead of bringing in among the plants usually termed "subtropical"
the fine hardy kinds which will be found so useful in the same way,
a special chapter is devoted to them; and in the case both of tender
and hardy plants, lists are furnished of such kinds as seed is procurable of, and which may be propagated in that way—the cheapest,
of course. It is needless to say any more in explanation of the
system, Mr. Gibson, the able superintendent of Battersea Park,
having so well shown the public of what superb results it is capable
by that not very large but inexhaustible garden, which he has so
cunningly and tastefully cut off from the rest of the park, and to a
great extent from the force of the breeze—often more injurious to
the large-leaved plants than the cold. The term "subtropical" is
not a very appropriate one, but we have here to deal only with
what the system has taught us, and in how far it is adoptable in
our gardens.

First, then, as to its teachings. It has taught us the value of
grace and verdure amid masses of low and brilliant and unrelieved
flowers, or rather reminded us of how far we have diverged from
Nature's ways of displaying the beauty of vegetation. Previous to
the inauguration of this movement in England, our love for rude
colour had led us to ignore the exquisite and inexhaustible way in
which plants are naturally arranged—fern, flower, grass, shrub, and
tree, sheltering, supporting, and relieving each other. We cannot
attempt to reproduce this literally, nor would it be wise or convenient to do so; but assuredly herein will be found the source of true
beauty in the plant world, and the more the ornamental gardener
keeps the fact before his eyes, the nearer truth and success will be
attained. Nature *in puris naturalibus* is not a beauty to be added
to our gardens, but Nature's laws should not be violated, and but
few human beings have contravened them more than our flower
gardeners during the past twenty years. We must compose from

Nature, as the best landscape artists do, not imitate her basely. We may have all the shade, the relief, the grace, and the beauty, and nearly all the irregularity of Nature seen in every blade of grass, in every sea-wave, and in every human countenance; and which may be found too, in some way, in every garden that affords us lasting pleasure either from its contents or design. Well, then, subtropical gardening has taught us that one of the greatest mistakes ever made in the flower garden was the adoption of a few varieties of plants for culture on a vast scale, to the exclusion of interest and variety, and too often of beauty or taste. We have seen how well the pointed, tapering leaves of the Canna carry the eye upwards; how refreshing it is to "cool the eyes" in the deep umbrage of those thoroughly tropical Castor-oil plants with their gigantic leaves; how grand the Wigandia, with its wrought-iron texture and massive outline, looks after we have surveyed brilliant lines and richly painted leaves; how greatly the sweeping palm-leaves beautify the British flower garden;—and, in a word, the system has shown us the difference between gardening that interests and delights all the public, as well as the mere horticulturist, and that which is too often offensive to the eye of taste, and pernicious to every true interest of what Bacon calls the " Purest of Humane pleasures."

But are we to adopt this system in its purity? Certainly not. All practical men see that to accommodate it to private gardens an expense and a revolution of appliances would be necessary, which are in nearly all cases quite impossible, and if possible, hardly desirable. We can, however, introduce to our gardens much of its better features; we can vary their contents, and render them more interesting by a cheaper and a nobler system. The use of all plants without any particular and striking habit or foliage, or other distinct peculiarity, merely because they are " subtropical," should be tabooed at once, as tending to make much work, and to return —a lot of weeds; for " weediness" is all that I can write of many Solanums and stove plants of no real merit which have been employed under the name subtropical. Selection of the most beauti-

ful and useful from the great mass of plants known to science is
one of the most important of the horticulturist's duties, and in no
branch must he exercise it more thoroughly than in this one.
Some plants used in it are indispensable—the different kinds
of Ricinus, Cannas in great variety, Polymnia, Colocasia, Uhdea,
Wigandia, Ferdinanda, Palms, Yuccas, Dracænas, and fine-leaved
plants of coriaceous texture generally.   A few specimens of
these may be accommodated in many large gardens; they will
embellish the houses in winter, and, transferred to the open garden
in summer, will lend interest to it when we are tired of the houses.
Some Palms, like Seaforthia, may be used with the best effect for
the winter decoration of the conservatory, and be stood out with an
equal result, and without danger in summer.   The many fine kinds
of Dracænas, Yuccas, Agaves, &c., which have been seen to some
perfection at our shows of late, are eminently adapted for standing
out in summer, and are in fact benefited by it.   Among the noblest
ornaments of a good conservatory are the Norfolk Island and other
tender Araucarias—these may be placed out for the summer much
to their advantage, because the rains will thoroughly clean and
freshen them for winter storing.   So with some Cycads and other
plants of distinct habit—the very things best fitted to add to the
present beauties of the flower garden.   Thus we may enjoy all the
benefits of what is called subtropical gardening without creating
any special arrangements for them in all but the smallest gardens.

But what of those who have no conservatory, no hothouses, no
means for preserving large tender plants in winter? They too may
enjoy in effect the beauty which may have charmed them in a subtropical garden.   We have no doubt whatever that in many places
as good an effect as any yet seen in an English garden from tender
plants, may be obtained by planting hardy ones only!   There is
the Pampas Grass—when well grown unsurpassed by anything that
requires protection.   Let us in planting it take the trouble to plant
and place it very well—and we can afford to do that, since one
good planting is all that it requires of us, while tender things of
one-tenth the value may require daily attention.   There are the

hardy Yuccas, noble and graceful in outline, and thoroughly hardy, and which if planted well, are not to be surpassed, if equalled, by anything of like habit we can preserve indoors.   There are the Arundos, conspicua and Donax, things that well repay for liberal planting; and there are fine hardy herbaceous plants like Crambe cordifolia, Rheum Emodi, Ferulas, and various fine umbelliferous plants that will furnish effects equal to those we can produce by using the tenderest.   The Acanthuses too, when well grown, are very suitable to this style; one called latifolius, which is beginning to get known, being of a peculiarly firm, polished, and noble leafage. Then we have a hardy palm—very much hardier too than it is supposed to be, because it has preserved its health and greenness in sheltered positions, where its leaves may not be torn to shreds by the storm through all our recent hard winters, including that of 1860.   And when we have obtained these we may associate them with not a few things of much beauty among trees and shrubs— with elegant tapering young pines, many of which, like Cupressus nutkaensis, have branchlets finely chiselled as a Selaginella, not of necessity bringing the larger things into close or awkward associa- tion with the humbler and dwarfer flowers, but sufficiently so to carry the eye from the minute and pretty to the higher and more dignified vegetation.   By a judicious selection from the vast mass of hardy plants now obtainable in this country, and by, where con- venient, associating with them house plants stood out for the summer, we may arrange and enjoy a beauty in the flower garden to which we are as yet strangers, simply because we have not suf- ficiently selected from and utilized the vast amount of vegetable beauty at our disposal.

Let us next select the finer tender plants for this purpose, to speak of the treatment they require, and the uses or associations for which they are best adapted.   In selecting tender plants of noble aspect or elegant foliage, suited for placing in the open air in British gardens during the summer months, we shall confine our- selves to first-class plants only.

It is necessary that they be such as will afford a distinct and beau-

tiful effect if they *do* grow, and that is by no means to be obtained
from many subjects recommended for "subtropical gardening."
And above all we must choose such as will make a healthy growth
in sheltered places in the warmer parts of England and Ireland at
all events. That some of the best will be found to flourish much
further north than is generally supposed there is some reason to
believe. In all parts the kinds with permanent foliage, such as the
New Zealand flax and the hardier Dracænas, will be found as effective as around Paris, and to such the northern gardener should turn
his attention as much as possible. Even if it were possible to cultivate the softer growing kinds like the Ferdinandas to the same perfection in all parts as in the south of England, it would by no means
be everywhere desirable, and especially where means are scarce, as
these kinds are not capable of being used indoors in winter. The
many find permanent leaved subjects that stand out in summer
without the least injury, and may be transferred to the conservatory
in autumn, there to produce as fine an effect all through the cold
months as they do in the flower garden in summer, are the best for
those with limited means.

ARALIA PAPYRIFERA (*the Chinese Rice-paper Plant*).—This, though
a native of the hot island of Formosa, flourishes beautifully around
Paris in the summer months, and is one of the most valuable plants
in its way. It is useful for the greenhouse in winter and the
flower garden in summer. It is handsome in leaf and free in
growth, though to do well it must be protected from cutting
breezes, like all the large-leaved things, and to this protection we
must pay due attention. In some of the warmer parts of France
the peach does very well as a field tree—a low one, however. The
wind is so strong that it would be destroyed if allowed to rise in the
natural way, and so they train it as a dwarf bush, spreading wide.
Tall "subtropical plants" have with us somewhat of the same dis-
advantage. If this Aralia be planted in a dwarf and young state,
it is like to give more satisfaction than if planted out when old and
tall. The lower leaves spread widely out near the ground, and then

it is presentable throughout the summer. Prefer therefore dwarf, stocky plants when planting it in early summer. It should have rich deep soil and plenty of water during the hot summer months. The open air of our country suits it better than the stove, and chiefly no doubt because it is very liable to the mealy bug when kept indoors—in the free air this pest is washed away by the rain. For the public gardens of Paris it is kept under ground in caves during the winter; but in private gardens with us it will doubtless be worthy of a place in the greenhouse throughout that season. It is

Fig. 2.—Aralia papyrifera.

easily increased by cuttings of the root. It is usually planted in masses, edged with a dwarfer plant; but as a small group in the centre of a bed of flowers, or even as an isolated specimen in a like position, it is capital. The stems of this plant have a very fine pure white pith, which, when cut into thin strips and otherwise prepared, forms the article known as rice paper. It is rare for a plant to be so useful both in an ornamental and economic sense.

ACACIA LOPHANTHA.—This elegant plant, though not hardy, is one of those which all may enjoy, from the freedom with which it grows in the open air in summer. It will prove more useful for the flower garden than it has ever been for the houses, and being easily raised is entitled to a place here among the very best. The elegance of its leaves and its quick growth in the open air make it quite a boon to the flower gardener who wishes to establish graceful verdure amongst the brighter ornaments of his parterre. It will furnish the grace of a fern, while close and erect in habit, thus enabling us to closely associate it with flowering plants without in the least shading them—except from ugliness. Of course I speak of its use in the young and single stemmed condition, the way in which it should be used. By confining it to a single stem and using it in a young state, you get the fullest size and beauty of which the leaves are capable. Allow it to become old and branched and it may be useful, but by no means so much so as when young and without side branches. It may be raised from seed as easily as a common bedding plant. By sowing it early in the year it may be had fit for use by the first of June; but plants a year old or so, stiff, strong, and well hardened off, for planting out at the end of May are the best. It would be desirable to raise an annual stock of this plant, as it is almost as useful for room decoration as for the garden.

CALADIUM ESCULENTUM.—This species has proved the best for out-door work of a large genus with very fine foliage. It is only in the midland and southern counties of Great Britain that it can be advantageously grown, so far as I have observed; but its grand outlines and aspect when well developed make it worthy of all attention, and of a prominent position wherever the climate is warm enough for its growth. It does very well about London, and may have been noticed in considerable masses in the London parks during the past year, and it served to illustrate the disadvantages of that mode of planting to some extent. When seen in wide masses the effect is by no means so fine as when in a compact group or circle. The dead level line presented by their tops—which, unlike that of the

upper surface of the taller plants, is below the eye—neutralizes considerably the great lines of the leaves ; but place the plant in a ring
round a central object, or in some position where its great leaves
may contrast immediately with those of a different type of vegetation, and it is beautiful indeed.

It may be used with good effect in association with many fine
foliage plants ; but Ferdinanda, Ricinus, and Wigandia usually grow
too strong for it, and if planted too close injure it.   This may

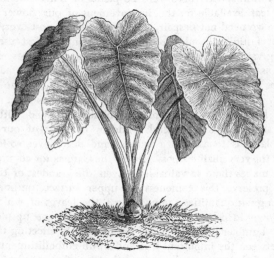

FIG. 3.—Caladium esculentum.

have been noticed particularly in cases where it was used as bordering to masses of the strong growing kinds above named.   With
all kinds of stonework, vases, &c., it is peculiarly effective and
beautiful. C. esculentum, though a stove perennial, is very easily
kept over the winter in a dry spot under a stage or in boxes of
sand in places where hot-house room is scarce.   It is readily propagated by first starting the plants in heat, and when they have

pushed forth eyes near the base, cutting them in pieces, an eye, or bud, in each.   In spring the older plants should be potted and grown on in heat, so as to be fit to plant out about the middle of June.   On the whole, although so fine and distinct, it is not suitable for any but mild and warm parts of the southern half of these islands; but in such is as indispensable as any other plant herein mentioned.

The Cannas.—If there were no plants of handsome habit and graceful leaf available for the improvement of our flower gardens but these we need not despair, for they possess almost every quality the most fastidious could desire, and present a most useful and charming variety.   The larger kinds make grand masses, while all may be associated intimately with flowering plants—an advantage that does not belong to some free-growing things like the Castor-oil plant.   The Canna ascends as boldly, and spreads forth as fine a mass of leaves as any; but may be closely grouped with much smaller subjects.   The general tendency of most of our flower-garden plants is to assume a flatness and dead level, so to speak, and it is the very qualities possessed by the Cannas for counteracting this that makes them so valuable.   Even the grandest of the other subjects preserve this tameness of upper surface outline when grown in great quantities : not so these, the leaves of which, even when grown in dense groups, always carry the eye up pleasantly from the humbler plants, and are grand aids in effecting that harmony between the important tree and shrub embellishments of our gardens and their surroundings, and the dwarf flower-bed vegetation which is so much wanted.

Another charm of these most useful subjects is their power of withstanding the cold and storms of autumn.   They do so better than many of our hardy open-air plants, so that when the last leaves have been blown from the lime, and the dahlia and heliotrope have been hurt from frost, you may see them waving as greenly and gracefully as the vegetation of a temperate stove.   Many of the subtropical plants, used for the beauty of their leaves, are so tender that they go off in autumn, or require all sorts of awkward protection at that season; but the Cannas last in good trim till the borders must be cleared. All sheltered positions, places near warm walls, and nice snugly-warmed dells, are capital positions for them.

Fig. 4.—Canna nigricans.

They are generally used in great ugly masses, both about Paris and London, but their true beauty will never be seen till we learn to place them tastefully here and there among the flowering plants—

just as we place sprigs of graceful fern in a bouquet.   A bed or two here and there solely devoted to them will, of course, prove very effective; but enormous meaningless masses of them, containing perhaps several hundred plants of one variety, are things to avoid and not to imitate.   As to culture and propagation nothing can be more simple : they may be stored in winter as readily as potatoes, under shelves in the houses, in the root-room, or, in fact, anywhere covered up from the influences of frost.   And then in spring, when we desire to propagate them, nothing is easier than pulling the roots in pieces, and potting them separately.   Afterwards it is usual to bring them on in heat, and finally harden them off previous to planting out; but some modification of this practice is desirable, as some kinds are of a remarkably hardy constitution, and make a beautiful growth if put out without so much as a leaf.

In rambling through an obscure part of Paris one evening, I encountered a tuft of Canna springing up strongly through and around a box-edging—pretty good evidence that it had remained there for some years.   Upon inquiry of the proprietor of the garden I found this was the case, and that he had no doubt of the hardiness of several other kinds.   How deep were they planted ?   Not more than eight or ten inches.   When we consider that the Cannas are amongst the most valuable plants we use for giving grace and verdure to the flower garden, this surely is worthy of being acted upon. Considering their diversity of colour and size, their graceful pointed habit and facility of propagation, we must concede them the first place, but their capability of being used by anybody who grows ordinary bedding plants, and the fact, that they may be preserved so very easily through the winter enhances their value still more. The following are among the best of the hardiest kinds :— C. Annæi, musæfolia, gigantea, limbata Warscewiczii, nigricans, and zebrina.   Of course they will prove equally hardy with us. As it is desirable to change the arrangements as much as possible every year, it may not be any advantage to leave them in the ground, and in that case they may be taken up with the bedding plants, and stored as simply and easily as carrots, parsnips, or potatoes. A bed of Cannas, protected by a coating of litter, was left out in Battersea Park through the severe winter of 1866-7. During the unfavourable summer of 1867 they attained a height of nearly twelve feet.

THE DRACÆNAS.—Long as this noble family have been known in our gardens we have yet to learn a great deal about its use and beauty. Hitherto allowed to grace a stove or conservatory now and then, for the future Dracænas will be among the most indispensable ornaments of every garden where grace or variety is sought. They are among the very best of those subjects which may be brought from the conservatory or greenhouses in early summer, and be placed in the flower garden till it is time to take them in again to the houses, where we protect them through the winter. And if it were not necessary to protect them through the winter it would be almost worth our while to bring them indoors at that season, so graceful are they, and so useful for adding the highest beauty to our conservatories. One well filled with such plants presents a very different appearance to what most English plant-houses do in winter. The hardier and most coriaceous kinds, like indivisa and Draco, may be placed out with impunity very far north. The brightly coloured kinds, like terminalis, have been tried in the open air at Battersea, but not with success. It would be dangerous to try them in the open air much further north, except in very favourable spots. The better kinds are indicated in the select list of subtropical plants. I have seen D. indivisa grow well in the open air in the south of England.

ECHEVERIA METALLICA.—This is scarcely elevated enough to be suitable for association with such plants as the foregoing, but it is so very distinct in aspect, and has been proved to grow so well in the open air during the past two unfavourable seasons, that we must not pass it by. I purposely exclude from this selection many things sometimes included in lists of "subtropical" plants, but which may be classed most properly with bedding subjects. But this, although not very large, forms an agreeable and distinct object, and is very well calculated for producing a striking effect among dwarf bedding and edging plants. It should, however, be placed singly, and among very dwarf things, such as Sedum, Sempervivum, and its dwarf relative E. secunda.   So arranged, it was very beautifully seen at Battersea last year; also charmingly done by Mr. Roger, at Bury Hill, Taplow ; and even so far north as Osberton, Notts, by Mr. Bennett. It may be propagated by the leaves or by cuttings, and requires a dry greenhouse shelf in the winter. Light sandy earth, not of necessity very poor, will suit it best in the open air.

FERDINANDA EMINENS.—This is one of the tallest and noblest subtropical plants, growing well in the southern and midland counties : wherever it is supplied with rich soil and abundant moisture. It is also very much the better of being sheltered, and so are all large and soft-leaved plants. Where the soil is rich, deep, and humid, and the position warm, it attains great dimensions, sometimes growing over twelve feet, and suspending immense pairs of opposite leaves.   It will in all cases form a capital companion to the Castor-oil plant, and though it may not be grown with such ease as that in all parts, it should be in every collection, growing quite as well in the south of England as in the neighbourhood of Paris. It requires to be planted out in a young state, and grows freely from cuttings. Greenhouse treatment will do in winter. It is better to keep a stock in pots through the summer to afford cuttings, though the old ones may be used for that purpose.

FICUS ELASTICA (*Indiarubber Plant*).—Another fine old plant, for which we have lately found a new use. It is one of those valuable leathery-leaved things that are useful in hothouse, drawing-room, or flower garden. It not only exists in the open air in summer in good health, but makes a good growth under the influence of our weak northern sun. Never assuming the imposing proportions of other plants mentioned here, it is best adapted for select mixed groups, and in small gardens as isolated specimens amongst low bedding plants. It requires stove treatment, and is propagated from cuttings. In all cases it is better to use plants with single stems. It is especially valuable in consequence of doing perfectly well in the dry air of inhabited rooms, and this ill

FIG. 5.—Ficus elastica.

enable many to enjoy a fine-leaved plant in the flower garden who have not a glass house of any kind on their premises.

MUSA ENSETE.—The noblest of all the plants yet used in the flower garden is *Musa Ensete*—the great Abyssinian Banana, discovered by Bruce. The fruit of this kind is not edible, unlike that of the Banana and Plantain (Musa paradisiaca and sapientum), but

the leaves are magnificent; and, strange to say, they stand the rain
and storms of the neighbourhood of Paris without laceration, while
all the other kinds of Musa become torn into shreds.   It is an interesting and hitherto unknown fact, that the finest of all the
Banana or Musa tribe is also the hardiest and most easily preserved.
When grown for the open air, it will of course require to be kept
in a house during winter, and planted out the first week in June.
In any place where there is a large conservatory or winter garden,
it will be found most valuable, either for planting out therein,
or for keeping over the winter, as, if merely housed in such a
structure during the cold months, it will prove a great ornament
among the other plants, while it may be put out in summer when
the attraction is all out of doors.   Other kinds of Musa have been
used with us, but have barely grown more in the open air than to
make it clear that they should not be so cultivated in this country.
This is the only species really worth growing for the open air. Where
the climate is too cold to put it out of doors in summer, it must yet
be grown in all conservatories in which it is desired to establish the
noblest type of vegetation. It has hitherto been generally grown in
stoves.   Not only are the leaves magnificent in their development,
but of a texture that seems . to withstand the heaviest rains and
storms.   The plant is difficult to obtain as yet, but will, I trust, be
sought out and made abundant by our nurserymen.   It is not generally known that this plant is of a remarkably hardy constitution,
and that it will grow well in a greenhouse or conservatory. Planted
out in a winter garden, it will grow healthfully, and I need not say
what a magnificent object it is for the decoration of such a place.
It also, strange to say, stands the drought and heat of a living room
remarkably well, and though, when well developed, it is much too
big for any but Brobdingnagian halls, the fact may nevertheless be
taken much advantage of by those interested in room decoration
on a large scale.

POLYMNIA GRANDIS, MACULATA, AND PYRAMIDALIS. — These
belong to the great composite order, and are distinguished by rich

and handsome foliage, and rapid summer growth, which, moreover, never becomes objectionable from any trace of raggedness, the erect shoots growing away till the end of the season in our climate. Doubtless, there is a point at which in their native country seediness does arrive, but with us they, like the Ricinus of one summer,

FIG. 6.—Polymnia grandis.

always look fresh and young, and are most appropriate for forming luxuriant masses of foliage in the flower garden, and for diversifying its aspect. P. grandis is best known in this country, and is second to no other plant for its dignified and yet finished effect in the flower garden; but P. maculata and P. pyrami-

c

dalis are also excellent and distinct. They are easily struck from
cuttings taken from old plants put in heat in spring, and are, like
most large soft growing things in this way, best planted out in a
young state, so as to insure a fresh and unstinted growth. P. pyramidalis is the newest of the group, and that least known in cultivation.
I saw it several times during the past season in Paris. The leaves
are not so large as those of the other species, and differ in shape,
being nearly cordate, but the growth is most vigorous and the habit
distinct. It pushes up a narrowly pyramidal head of foliage to a
height of nearly ten feet in Paris gardens, and will be found to do
well in the south of England.

PHORMIUM TENAX (*the New Zealand Flax*).—This is tolerably
well known among us as a greenhouse and conservatory subject,
but not nearly so much grown as it ought to be. What a grand
use the French make of it, both indoors in the winter, and in
the conservatory and out of doors in summer! About Paris it is
of course as tender as with us, and requires the same amount of
attention, which, after all, is very little. They grow it by the
thousand for the decoration of rooms, and in the great nursery
of the city of Paris at Passy there are 10,000 plants of it, chiefly
used for the embellishment of the Hôtel de Ville. I need hardly
say that we are much worse off for graceful things for indoor decoration than the French, and should in consequence grow this
plant abundantly, according to our space. When grown to a
medium size its leaves begin to arch over, and when in that condition nothing makes a more graceful and distinct ornament for
room or hall. It may be grown to presentable perfection in an
eight-inch pot, or to a great mass of bold long leaves in a tub a yard
in diameter. Generally with us it will be found to enjoy greenhouse
temperature, though in genial places in the south and west of
Ireland and England it does very well in the open air. Its best
use is for the decoration of the garden in summer, a few specimens well grown and plunged in the grass or the centre of a bed
giving a most distinct aspect to the scene. The larger such

plants are the better the effect, of course. The smaller plants will prove equally useful and effective in vases, to which they will add a grace that vases rarely now possess. It is pre-eminently useful from its being alike good for the house, conservatory, and even the living rooms in winter. Wherever indoor decoration on a large scale is practised it is indispensable, and it should be remarked that, unless for vase decoration, it requires to be grown into goodly specimens before affording much effect out of doors; but when grown large in tubs, it is equally grand for the large conservatory and for important positions in the flower garden.

RICINUS COMMUNIS (*the Castor-oil Plant*).—When well grown in the open air, there is not in the whole range of cultivated plants a more imposing subject than this. It may have been seen nearly twelve feet high in the London parks, and with leaves nearly a yard wide. It is true we require a bed of very rich deep earth under it to make it attain such dimensions and beauty; but in all parts, and with ordinary attention, it grows well. In warm countries, in which the plant is very widely cultivated, it becomes a small tree, but is much prettier in the state in which it is seen with us—*i.e.*, with an unbranched stem, clothed from top to bottom with noble leaves. Soon after it betrays a tendency to develope side-shoots, the cold autumn comes and puts an end to all further progress; and so much the better, because it is much handsomer in a simple-stemmed state than any other. And the same is true of not a few other large-leaved plants—once they break into a number of side-shoots their leaf beauty is to a great extent lost. In the planting out of some other subjects, it has been considered well to raise the beds on lime-rubbish, &c., or in other words, building them upon it, sloping up the edge with soil and turf. But to grow this to perfection, the best way is to deeply excavate the bed, and place some rich stuff in the bottom, making all the earth as rich as possible. It is as easily raised from seed as the common bean, requiring, however, to be raised in heat. The Ricinus is a grand plant for making bold and noble beds near those of the more brilliant flowers, and tends to finely vary the flower garden. It is not well to closely associate it with bedding plants, in consequence of the strong growth and shading power of the leaves, so to speak. To make a compact group of the plant in the centre of some wide circular bed, surrounded with a band of a dwarfer subject, say the Aralia or Caladium, and then whatever arrangement may be most admired of the flowering plants, is a good plan—a bold and magnificent centre is obtained, while the effect of the flowers is much enhanced, especially if the planting be nicely graduated and tastefully done. It is a judicious combination of both the green and the gay that we are most in want of, and few things can do so much to effect it for us in the flower garden as the common Castor-oil plant. This combination may, and must be, effected in any way that taste may direct. A graceful handsome-leaved subject in the centre of a flower bed will help it out, and so will bold groups of fine-leaved plants towards the outer parts of the flower garden. These bold masses connect in some degree the larger ligneous vegetation that usually surrounds our flower gardens with the small and low-lying brilliant flowers. For such groups the varieties of the Castor-oil plant are not likely to be surpassed.

SEAFORTHIA ELEGANS.—This is perhaps the most elegant and useful of all palms which may be safely placed out in summer. It is too scarce as yet to be procurable by horticulturists generally, but should be looked for by all who take an interest in these matters, and have a house in which to grow it. It stands well in the conservatory during the winter, though generally kept in the stove, where of course it grows beautifully. There are hardier kinds—the dwarf fan palm for example, but on the whole none of them are so valuable as this. The following palms are suitable for like purposes :—

Areca lutescens.	Chamærops Palmetto.
Caryota urens.	Latania borbonica.
„ sobolifera.	Phœnix dactylifera.
Chamærops humilis.	„ sylvestris.
Fortunei.	Corypha australis.

THE SOLANUMS.—This family, so wonderfully varied, affords numerous species that look fine and imposing in leaf when in a young and free-growing state. In the nursery garden of the city of Paris there is, entirely devoted to the family, a very large house in which are preserved over the winter months more than sixty species for the embellishment of Parisian gardens. But in selecting from this great genus we must be much more careful, as our climate is a shade

FIG. 7.—Solanum Warscewiczii.

too cold for them, and many of them are of too ragged an aspect to be tolerated in a tasteful garden. Half-a-dozen or more species are indispensable, but quite a crowd of narrow-leaved and ignoble ones may well be dispensed with. The better kinds—as seen both in London and Paris gardens—are marginatum, robustum, macranthum, macrophyllum, Warscewiczii, crinipes, callicarpum, jubatum, Quitoense, galianthum, hippoleucum, crinitum, and Fontainesianum, an annual with pretty leaves, crisped and distinct looking.

Most of these plants may be raised from seed, while they are also freely grown from cuttings. As a rule hothouse treatment in winter is required, and in summer rich light soil, a warm position, and perfect shelter. S. marginatum planted in a very

Fig. 8.—Solanum robustum.

dwarf and young state, furnishes a most distinct and charming effect: it should be planted rather thinly, so that the leaves of one plant may not brush against those of another. If some very dwarf plants are used as a groundwork so much the better, but the downy and silvery leaves of this plant are sure to please without this aid. It is very much better when thus grown than when permitted to assume the bush form. All the other kinds named are suitable for association with the larger leaved plants, though they do not attain such height and vigorous development as those of the first rank, like the Ricinus.

UHDEA BIPINNATIFIDA.—This is one of the most useful plants in its class, producing a rich mass of handsome leaves, with somewhat

FIG. 9.—Uhdea bipinnatifida.

the aspect of those of the great cow-parsnips, but of a more refined type. The leaves are of a slightly silvery tone, and the plant continues to grow fresh and vigorously till the late autumn. It is freely propagated by cuttings taken from old plants kept in a very cool stove, greenhouse, or pit during the winter months, and placed in heat to afford cuttings the more readily in early spring. Under ordinary cutting treatment on hotbeds or in a moist warm propagating house, it grows as freely as could be desired, and may be planted out at the end of May or the beginning of June. It is well suited for forming rich masses of foliage, not so tall, however, as those formed by such things as Ricinus or Ferdinanda.

VERBESINA GIGANTEA, *and other species.*—To these somewhat the same remarks will serve as have been applied to the preceding. They require about the same treatment, and are useful in the production of like effects. They, like their fellows, will be much the better for as warm and sheltered a position and as rich and light a soil as can be conveniently given them.

WIGANDIA MACROPHYLLA (*caracasana*).—This noble plant, a native of the mountainous regions of New Granada, is unquestionably, from the nobility of its port and the magnificence of its leaves, entitled to hold a place among the finest plants of our gardens.. Under the climate of London it has made leaves which have surprised all beholders, as well by their size as by their strong and remarkable veining and texture. There can be little doubt that it will be found to succeed very well in the Midland and Southern counties of England, though too much care cannot be taken to secure it a warm sheltered position, free good soil, and perfect drainage. It may be used with superb effect either in a mass or as a single plant. It may be raised from seed, and seed is offered of it in some of our catalogues. W. urens is a relative, not so good by any means; and W. Vigieri has been recently added to our collections, and given a very good character. Some have even asserted it to be better than W. macrophylla, but that is not the case. So far as I have observed it makes a respectable growth in the neighbourhood of Paris, but never assumes the majestic aspect

of W. macrophylla when well grown ; but being readily raised from
seed, it is worth a place.   Seeds of the three species are offered in the
seed lists for the present year.   W. macrophylla, the most valuable
species, is frequently propagated by cuttings of the roots and grown
on in a moist and genial temperature through the spring months,

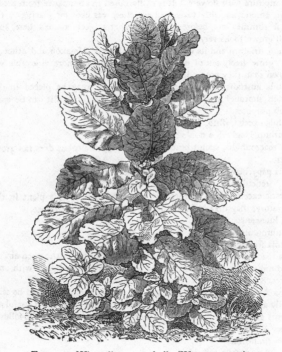

FIG. 10.—Wigandia macrophylla (W. caracasana).

keeping it near the light so as to preserve it in a dwarf and well
clothed condition; and, like all the other plants in this class, it
should be very carefully hardened off.   It is, however, much better
raised from cuttings of the shoots, if these are to be had.

A select list, with notes, of 100 *of the subtropical plants best suited for use in our climate. The most indispensable kinds are marked* *.

1. *Acacia lophantha, does freely in open air, and is very graceful and suitable for mixture with flowers. It is easily raised in abundance from seed.

2. *Agave americana with variegated varieties, was used for placing in the open air in summer long before any of the other subjects named here, and when well grown looks very imposing.

3. Abutilon striatum and its varieties, such as Duc de Malakoff and others, flower and grow freely out of doors in summer, and are more enjoyable when so grown than they are under glass.

4. Alsophila australis, a tree and large stove fern, may be placed in the open air in thoroughly well-sheltered and shady positions if it can be spared for such purposes.

5. Alsophila excelsa, ditto.

6. Anthurium Hookeri, a noble species; stands well.

7. Aralia macrophylla, stands beautifully in the open air, but does not grow freely there.

8. *Aralia papyrifera, one of the best and freest of all.

9.   „   reticulata.

10. Araucaria excelsa. Where there are nice specimens of this plant in the conservatory, they may be used out in summer with much taste.

11. Areca lutescens.

12. Balantium culcitum.

13. Bocconia frutescens.

14. Brexia madagascariensis, a good plant, and stands well in the open air.

15. *Caladium esculentum. One of the best; fine for association with sculpture, vases, &c. Likes rich soil, warmth, and shelter.

16. *Canna Annei superba. Of this and the other Cannas little need be said, especially as they are alluded to elsewhere. It is by far the most useful group of plants of all used in this way. Among the better kinds are the following:—

17. *Canna Annei limbata.

18.   „   *robusta.

19.   „   *musæfolia hybrida.

20.   „   *nigricans.

21.   „   *grandiflora floribunda.

22.   „   *Géant.

23.   „   *discolor floribunda.

24.   „   *metallica.

25.   „   *rubra superbissima.

26. Carludovica palmata.
27. Caryota urens.
28. „ sobolifera.
29. Cassia corymbosa, flowers pretty freely, but is scarcely distinct enough to be generally admired.
30. Cassia floribunda, ditto.
31. Chamæpeuce diacantha, a plant of very striking aspect.
32. Chamærops humilis.
33. „ *excelsa.
34. „ *Palmetto.
35. Colea Comersoni, a plant with very noble leaves, as yet rare.
36. *Colocasia odorata, when old looks peculiarly distinct and magnificent, in consequence of being elevated on a bold stem.
37. Cordyline indivisa. This fine thing will stand out of doors without injury in summer, even in the far north.
38. Corypha australis.
39. Cyathea dealbata. For use where other tree ferns are ventured out.
40. *Cycas revoluta, as useful in the conservatory as in a flower garden; will embellish the first through the cold months, and form a distinct and graceful subject in the centre of a choice bed of flowers all the summer.
41. Dahlia imperialis.
42. Dicksonia antarctica. Probably the best of the tree ferns for placing in the open air.
43. *Dracæna australis. The Dracænas here given are all fine, and very useful in their way.
44. *Dracæna indivisa.
45. „ *Draco.
46. „ *braziliensis.
47. „ nutans.
48. „ Rumphi.
49. „ erithorachis.
50. „ *cannæfolia.
51. „ *lineata.
52. *Echeveria metallica, very distinct in colour and appearance, and grows well out of doors in England.
53. Erythrina crista-galli, and its varieties, flower well in the south of England, as well as around Paris.
54. *Ferdinanda eminens, one of the best and most popular.
55. *Ficus elastica, never suffers, but makes free growth in the open air, and is useful at all seasons, indoors or out.
56. Ficus nympheæfolia, scarce in private collections, but very fine.
57. Grevillea robusta.

58. Hedychium aurantiacum, a well known old stove plant, which, placed in a favourable position, opens its flowers freely in the open air; that, however, does not make it worth growing for this purpose, except in large collections.

59. Hedychium Gardnerianum, ditto.

60. Lomatia Bidwilli, has finely dissected and handsome foliage.

61.   ,,   silaifolia, ditto.

62. Lomatophyllum borbonicum.

63. *Melianthus major, of very handsome foliage and agreeable glaucous colour. Out of doors in a dry place it gets cut down annually by frost, but comes up again in spring, and makes a very presentable show of dwarf and well-clothed shoots, and is in fact much handsomer when seen in this state than when grown indoors. It is better when left out of doors in winter and slightly protected, than when turned out annually from the houses.

64. Monstera deliciosa, a fine stove fruit; may be placed in the open air in summer with impunity.

65. *Musa Ensete, the great Musa of Abyssinia, discovered by Bruce; the noblest of all subtropical plants.

66. Neottopteris australasica.   Striking fern for placing in shady nooks.

67. Nicotiana Wigandioides.

68. Papyrus antiquorum.

69. Philodendron Simsi, stands well out in England.

70.   ,,   macrophyllum, ditto; is a very fine plant.

71. Phœnix dactylifera.

72.   ,,   sylvestris.

73. *Phormium tenax, a most useful plant for this purpose, and for every purpose of ornament, whether indoors or out.

74. *Polymnia grandis, also first rate.

75.   ,,   maculata, ditto.

76. Pothos acaulis, suffers a little, and should be placed in a well-sheltered and warm place.

77. Rhopala corcovadense, one of a type of fine-foliaged stove plants that may be placed out of doors in summer in a sheltered spot.

78. *Ricinus communis, in many varieties, is perhaps on the whole the most useful and easily raised of all subtropical plants except the Cannas.

79. Sanseviera zeylanica.

80. *Seaforthia elegans, a palm which stands our summer climate without injury, and is of course among the most useful of ornamental plants indoors at all other seasons.

81. Selinum decipiens, an umbelliferous plant, of distinct character and handsome leaves.

82. Senecio Ghiesbreghti.

83. Senecio Petasites.
84. Solanum crinipes.
85.   „   macranthum.
86.   „   macrophyllum.
87.   „   marginatum.
88.   „   robustum.
89.   „   Warscewiczii.
90. Sonchus laciniatus.
91. Sparmannia africana, produces its singularly pretty flowers freely in the open air.
92. Stadmannia Jonghii.
93. Tradescantia discolor, very useful as a bordering plant to beds of subtropical flowers, from the attractive colouring of the backs of its leaves.
94. Tradescantia zebrina.
95. Tupidanthus calyptratus, grows a little in the open air, and stands well.
96. *Uhdea bipinnatifida, a free-growing, large-leaved, and very useful plant for this purpose.
97. Verbesina gigantea.
98.   „   *verbascifolia.
99. *Wigandia caracasana, one of the noblest of fine-leaved plants when grown in the flower-garden in summer.
100. Wigandia Vigieri.

List of the best twenty-four Subtropical Plants.

1. Acacia lophantha.	13. Ficus elastica.
2. Agave americana.	14. Melianthus major.
3. Aralia papyrifera.	15. Musa Ensete.
4. Caladium esculentum.	16. Phormium tenax.
5. Canna Annei superba.	17. Polymnia grandis.
6. Chamærops excelsa.	18. Ricinus communis.
7. „ humilis.	19. Seaforthia elegans.
8. Cordyline indivisa.	20. Solanum marginatum.
9. Cycas revoluta.	21. „ Warscewiczii.
10. Dracæna Draco.	22. Uhdea bipinnatifida.
11. „ indivisa.	23. Verbesina gigantea.
12. Ferdinanda eminens.	24. Wigandia caracasana.

Subtropical Plants that may be raised from Seed.

Abutilon, in variety.	Andropogon Sorghum.
Acacia lophanta.	Aralia australis.
Andropogon bombycinus.	„ elegans.
„ formosus.	„ papyrifera.

Subtropical Gardening.

Aralia Sieboldi.
„ trifoliata.
Areca sapida.
Artemesia argentea.
Bambusa himalaica.
Bocconia cordata.
„ formosa.
„ frutescens.
„ japonica.
„ macrophylla.
Brugmansia, in variety.
Canna, in profuse variety.
Cassia corymbosa.
„ floribunda.
Chamæpeuce Cassabonæ.
„ diacantha.
Chamærops humilis.
„ „ glauca.
„ macrocarpa.
Cineraria platanifolia.
Cordyline indivisa vera.
„ nutans „
, superbicus.
„ Veitchii.
Corypha australis.
Cyperus vegetus.
Dahlia imperialis.
Erianthus Ravennæ.
„ violaceus.
Erythrina caffra.
„ crista galli.
„ Hendersoni.
„ laurifolia.
Eucalyptus globulus.
Ferdinanda eminens.
Grevillea robusta.
Hedychium Gardnerianum.

Humea elegans.
Latania borbonica.
Melianthus major.
„ minor.
Musa Ensete.
Nicotiana grandiflora, a variety of N. tabacum.
Nicotiana Wigandioides.
Owenia cerasifera.
Paratropia tomentosa.
„ venulosa.
Phormium tenax.
Phytolacca dioica.
Polymnia grandis.
Ricinus, in variety.
Seaforthia elegans.
Solanum acanthocarpum.
„ auriculatum.
„ giganteum.
„ glaucophyllum.
„ glutinosum.
„ lanceolatum.
„ macrocarpum.
„ macrophyllum.
„ marginatum.
„ pyracanthum.
„ robustum.
„ verbascifolium.
Sonchus pinnatus.
Sparmannia africana.
Uhdea bipinnatifida.
Verbesina verbascifolia.
Wigandia macrophylla.
„ 'urens.
„ vigieri.
Zea japonica variegata, and others.

All the above have been offered in Seed Catalogues of the current year.

CHAPTER II.

Hardy Plants for "Subtropical" Gardening.

THE taste for varying the surface of the flower garden, and for adding to its interest and character by using plants of a nobler type than we have already generally done, is sufficiently developed to make it desirable that we should search out the most desirable hardy plants for such purposes. There are a great many tender plants recommended, a great many used; but, as usual, the really meritorious are in a minority. In many parts it is a dangerous and useless thing to place tender plants in the open air for the sake of producing such an effect as may have been seen at Battersea. Some have placed out things that have no chance of looking to any advantage out of a hothouse in this country—for instance, all the Bananas except M. Ensete. Only such things should be used as will stand our summer climate without injury, or grow freely and luxuriantly in it. Of course, hardy plants are the very best of all, but as yet we are far from having a sufficiency of these of the precise type that is required, and we rarely make good use of some fine subjects pretty well known. Nothing is more lamentable than to see a flower garden with tender plants perishing from cold in the middle of summer. The climate and altitude of the garden will frequently prevent the use of tender things, for it is well known that you may grow many things in the open air about London and southwards which will barely exist, if at all, in the north. Therefore, the position and capabilities of a garden are the first things that should be taken into consideration. It by no means follows that because a place proves cold and uncongenial to many tender subjects which have afforded a good effect about London, that we should rest satisfied with the usual tame aspect of things. There is no part in these islands for which fine, verdant, and distinct-looking hardy flower-garden plants will not eventually be found, if they be not in the country already.

More than that, we may produce as good effects from the use of hardy plants alone as any afforded by what are called "subtropical,' and in the following pages an attempt is made at pointing out the most suitable kinds to use. It is a most important subject, and concerns every gardener in the British Isles; for however few can indulge in the luxury of rich displays of tender plants, or however rare the spots in which they may be ventured out with confidence, all may enjoy those that are hardy, and that too with infinitely less trouble than is required by the tender ones. Those noble masses of fine foliage first displayed to us in this country by the able and energetic superintendent of Battersea Park, have done much towards correcting a false taste. What I wish to impress upon the reader is, that let him live where he may in these islands he need not despair of producing enough of like effect to beautifully vary flower garden or pleasure ground *by the use of hardy plants alone;* and that the noble lines of a well-grown Yucca recurva, or the finely-chiselled, yet graceful fern-like spray of a graceful young conifer, will aid him as much in this direction as anything that grows either in tropical or subtropical climes.

The herbaceous collections in the Jardin des Plantes are very full, and correctly kept; and being much devoted to such plants, I rarely spent a week without visiting them, chiefly to discover useful plants in this class; while, of course, such as are used in the various public parks and gardens also came repeatedly under my observation. Of their relative importance and value I was, perhaps, the better prepared to judge from having visited for like purposes all the botanic gardens in the British Isles within the past few years. The following are the finest subjects we can use :—

ACANTHUS LATIFOLIUS.—This is a plant that anybody can grow, and which is in all respects fine. The leaves are bold and noble in outline, and the plant has a tendency, rare in some hardy things with

FIG. 11.—Acanthus latifolius.

otherwise fine qualities, to retain its leaves till the end of the season without losing a particle of its freshness and polished verdure. In fact the only thing we have to decide about this subject is, what is the best

D

place for it? Now, it is one of those things that will not disgrace any position, and will prove equally at home in the centre of the mixed border, projected a little from the edge of a choice shubbery in the grass, or in the flower garden; nobody need fear its displaying anything like the seediness which such things as the Heracleums do at the end of summer. In fact, there are few things turned out of the houses that will furnish a more satisfactory effect. I should not like to advise its being planted in the centre of a flower bed, or in any other position where removal would be necessary; but in case it were determined to plant permanent groups of fine-leaved hardy plants, then indeed it could be used with great success. Supposing we have an irregular kind of flower garden or pleasure ground to deal with (a common case everywhere), one of the best things to do with it is to plant it in the grass, at some little distance from the clumps, and near perhaps a few other things of like character. It is better than any kind of Acanthus hitherto commonly cultivated in botanic gardens, though one or two of these are fine. Give it deep good soil, and do not begrudge it this attention, because, unlike tender plants, it will not trouble you again for a long time. How about a ring of it around a strong clump of Tritomas (grandis in the middle, and glaucescens surrounding it?) the very dark polished green Acanthus being in its turn surrounded by that fine autumn-flowering Sedum spectabile? There would be little difficulty in suggesting a dozen equally suitable uses for this fine plant. It is to be had now in some London nurseries, and in nearly every Paris one. The plant best known by this name as also known under the name of A. lusitanicus. Both are garden names, the first the best. It came into cultivation in the neighbourhood of Paris about six or eight years ago, and has since spread about a good deal. Nobody seems to know from whence it came. Probably it is a variety of Acanthus mollis—a striking one, of course. The plant itself varies a good deal; I have seen specimens of it about a foot high, with leaves comparatively small and stiff and rigid, as if cast in a mould, by the side of others of thrice the development, and of the usual texture.

ANDROPOGON SQUARROSUM is a hardy plant in the neighbour-hood of Paris, or survives with but slight protection, making luxuriant tufts seven feet high or more, when in flower. It would probably make a beautiful object in the warmer and milder parts of England and Ireland, and in good soil, but, unlike the preceding, it is not a subject which can with confidence be recommended for every garden. But all who value fine grasses should try it.

ARALIA EDULIS (*Sieb. and Zucc.*)—This is a vigorous plant, highly suited for adding distinction to those positions in which we desire a luxurious type of vegetation. It is perfectly hardy, grow six, seven, and even eight feet high in good soil, is of a fresh and luxurious habit, and this so early as the end of June. The leaves attain a length of nearly a yard when the plant is strong, while the shoots droop a little with their weight, and thus it acquires a slightly weeping character. It is a little curious that plants so famous for their medicinal or other uses as the Castor-oil the Chinese rice-paper, and the Indian-rubber plants, should have be-come so very useful in the flower garden. For this edible Aralia we may claim as high a position as a hardy plant. For planting singly few things surpass it. It is very rare in this country now, but being easily propagated, may, it is to be hoped, not long prove so. I have seen it nine feet high; but as it dies down rather early in autumn it must not be put in important groups, but rather in a position where its disappearance may not be noticed. An isolated place, or one near the margin of an irregular shrubbery, fernery, or rough rockwork by the side of a wood walk, will best suit.

ARALIA JAPONICA.—A frutescent species, and fine plant for varying the garden, bearing immense and graceful leaves, and delighting most in a warm and sheltered position—plenty of sun, but little exposure to wind. It is best when the stem is rather short and simple, and has an advantage that few things of the kind have—

it may be used with a stem of considerable height, or with a very dwarf one.

ARALIA SPINOSA, the angelica tree of North America, is highly useful in this class, in consequence of its beauty of foliage and distinct aspect. It, like many of the hardy things, should be placed in positions where it would not be necessary to remove it, or closely associate it with tender plants requiring frequent disturbance of soil. Both this and the preceding kind may be had in our nurseries.

ARALIA SIEBOLDI is also a valuable species, usually treated as a

FIG. 12.—Aralia Sieboldi.

greenhouse plant. It has been known to stand severe winters and afterwards grow better than plants that had been housed.

ARUNDO DONAX, the great reed of the south of Europe, is a very noble plant on good soils. In the south of England it forms canes ten feet high, and furnishes a very distinct and striking aspect.

It will do more than that if put in a rich deep soil in a favoured locality; and those who so plant clumps of it on the turf in their pleasure grounds will not be disappointed at the result. Nothing can be finer than the aspect of this plant when allowed to spread out into a mass on the turf of flower garden or pleasure ground. It seems to much prefer dry sandy soils to moist ones; indeed, I have known it refuse to grow on heavy clay soil, and flourish most luxuriantly on a deep sandy loam in the same district. It, like all large-leaved plants, loves shelter. No garden or pleasure ground in the southern parts of England and Ireland should be without a tuft of it in a sheltered spot. But, fine as it is for effect and distinctness, its variegated variety is of more value to the flower garden proper.

ARUNDO DONAX VERSICOLOR.—This is a wonderfully effective and beautiful plant, that is made little or no use of. We have already noticed several fine things for grouping together, or for standing alone on the turf and near the margin of a shrubbery border; and this is as well suited for close association with the choicest bedding flowers as an Adiantum frond is with a bouquet. It will be found hardy in the southern counties; and considerably north of London may be saved by a little mound of cocoa-fibre, sifted coal ashes, or any like material that may be to spare. In consequence of its effective and beautiful variegation, it never assumes a large development, like the green or normal form of the species, but keeps tidy and low, and yet thoroughly graceful and effective. It is of course suited best for warm, free, and good soils, and abhors clay, though it is quite possible to grow it even on that with a little attention to the preparation of the ground. But it is in all cases better to avoid things that will not grow freely and gracefully on whatever soil we may have to deal with; and it is to those having gardens on good sandy soils, and in the warmer parts of England, that I would specially recommend this grand variegated subject. For a centre to a circular bed, nothing can surpass it in the summer and autumn flower garden, while of course many other charming uses

may be made of it, not the least happy of which would be to plant
a tuft of it on the green grass, in a warm spot, near a group of
choice shrubs, to help, with many other things named, to fill up
the gap between ordinary fleeting flowers, and the taller shrub
and tree vegetation that is now nearly everywhere observed. It is
better to leave the plant in the ground, in a permanent position,
than to take it up annually. Protect the roots in the winter,
whether it be planted in the middle of a flower bed or planted by
itself in a little circle on the grass.

ARUNDO CONSPICUA is a worthy companion for the Pampas,
though by no means equal to it, as has been stated by some
writers. As a conservatory subject nothing surpasses it, and it will
be seen often in large conservatories after a few years. A large pot
or tub will be necessary if grown indoors. The drooping leafage
will always prove graceful, and then it sends up long silvery plumes,
drooping also, and strikingly beautiful. Judging by its different
appearance when freely grown in a tub indoors, and when planted
out even on favourable spots, my impression is that it by no means
takes so kindly to our northern climate as the Pampas grass. However, it is well worth growing, even in districts where it does not
attain great development; it comes in flower before the Pampas,
and may be considered as a sort of forerunner of that magnificent
herb.

BAMBUS VIRIDIS-GLAUCESCENS, *and others.*—I wish to call the
attention of all horticulturists who live in the southern and more
favoured parts of these islands to the fact that there are several
bamboos and bamboo-like plants from rather cool countries that
are well worth planting. Nothing can exceed the grace of a
bamboo of any kind, if freely grown; but if starved in a dirty hothouse, or grown in a cold dry place, where the graceful shoots
cannot arch forth in all their native beauty, nothing can be more
miserable in aspect. On cold bad soils, and exposed dry places in the
British Isles, these bamboos have no chance; but, on the other hand,

they will be found to make most graceful objects in many a sheltered nook in the south and south-western parts of England and

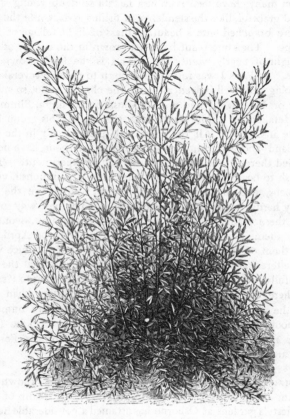

FIG. 13.—Bambusa aurea.

Ireland. Nowadays there is a growing taste for something else
than mere colour in the flower garden, and these will in many cases

be found a graceful help.   We have some knowledge of the capabilities of one kind in this country.   In a well sheltered moist spot
at Bicton many have seen Bambusa falcata send up young shoots,
long and graceful, like the slenderest of fishing-rods, while the older
ones were branched into a beautiful mass of light foliage of a distinct type.   The same plant has been grown in the county of Cork
to a height of nearly twenty feet.   This is the best known kind
we have.   At Paris I was fortunate enough to observe several other
kinds doing very well indeed, although the climate is not so suitable
as Cork or Devon.   These are Bambusa aurea, nigra, Simmonsii,
mitis, Metake, and viridis-glaucescens, the last of this group being
very free and good.   All the others will prove hardy in the south
of England and Ireland, though, as some of them have not yet
been tried there, it requires the test of actual experiment.   Those
who wish to begin cautiously had better take B. Simmonsii, viridis-
glaucescens, and nigra to commence with, as they are the most
certainly hardy, so far as I have observed.   The best way to treat
any of these plants, obtained in summer or autumn, would be
to grow them in a cool frame or pit till the end of April, then
harden them off for a fortnight or so, and plant out in a nice warm
spot, sheltered also, with good free soil—taking care that the roots
are carefully spread out, and giving a good free watering to "settle"
all.   There are no plants more worthy of attention than these
where the climate is at all favourable, and there are numerous
moist nooks around the British Isles where they will be found
to grow most satisfactorily.   The pretty little Bambusa Fortunei
is also hardy.

CHAMÆROPS EXCELSA.—It may not be generally known that
this palm is perfectly hardy in this country.   A plant of it in
her Majesty's gardens at Osborne has attained a considerable height.
It is also out at Kew, though protected in winter.   On the water
side of the high mound in the Royal Botanic Gardens, Regent's-
park, it is in even better health than at Kew, though it has not had
any protection for years, and has stood the fearfully hard frost of

1860.   If small plants of this are procured, it is better to grow them on freely for a year or two in the greenhouse, and then turn out in April, spreading the roots a little and giving deep loamy soil. Plant in a sheltered place, so that the leaves may not be injured by winds when they get large and grow up.   A gentle hollow, or

FIG. 14.—Chamærops excelsa.

among shrubs on the sides of some sheltered glade, will prove the best places.   The establishment of a palm among our some-what monotonous shrubbery and garden vegetation is surely worthy of a little trouble, and the precautions indicated will prove quite sufficient.

CRAMBE CORDIFOLIA.—This is unquestionably one of the finest of perfectly hardy and large-leaved herbaceous plants. It is as easily grown as the common seakale—easier, if anything; and in heavy rich ground makes a splendid head of leaves, surmounted in summer by a dense spray of very small flowers. Doubtless, if these be pinched off, a larger development of the fine glossy leaves may be expected, but as the shoots are so vigorously shot up and converted into a distinct and pretty inflorescence, many will prefer to "leave the plant to Nature." In planting it, the deeper and richer the soil, the finer the result. It will prove a capital thing for every group of fine-leaved hardy plants, and may also be popped in one or two places where a bold though low type of vegetation is desired. There is another species, C. juncea, which is also effective, but not so valuable as C. cordifolia.

CUCUMIS PERENNIS (*the Perennial Cucumber*).—This has not the quality of leaf which we could desire, but it will prove interesting to many. It is perfectly hardy, and possesses, so to speak, great trailing power. Its leaves are strong, rough, and of a glaucous colour; and the shoots run about freely if the plant be in very rich soil. Where bold trailing plants for high trellis-work, or rough banks, or shaggy rockwork are desired, it will be found distinct; but withal we cannot give it a place in the front rank, and the small select garden without any of the above-mentioned appendages will certainly be better without it. For the botanical garden and curious collections it is indispensable. It is strong and lasting when well established, and may be allowed to fall over rough banks, stumps, or be trained up trellis-work, &c.

DATISCA CANNABINA.—The male plant of this has long been known as a very strong and effective herb—graceful too; but I saw female plants associated with males for the first time in the Jardin des Plantes, and since then I have a higher opinion of the species. The female plant remains green much longer than the male, and being profusely laden with fruit, each shoot droops and the whole plant improves in aspect. It must not be forgotten in any selection of hardy plants of free growth and imposing aspect. From seed will probably be found the best way to raise it, and then one would be pretty sure of securing plants of both sexes.

Elymus arenarius.—This British grass—this strong-rooting and most distinct-looking herb—is capable of adding a striking feature to the garden here and there, and should be quickly introduced to civilization. Planted a short distance away from the margin of a shrubbery, or on a bank on the grass, and allowed to have its own way in deep soil, it makes a most striking object. In short, it deserves to rank fourth among really hardy big grasses, the Pampas and the two Arundos alone preceding it. I am not quite certain that it is not more useful than the Arundo, being hardy in all parts of these islands. In very good soil it will grow four feet high, and as it is for the leaves we should cultivate it, if the flowers are removed they will be no loss. It is found frequently on our shores, but more abundantly in the north than in the south. The variety called geniculatus, which has the spike pendulous, is also worthy of culture, and in its case the flowers may prove worth preserving. It may possibly be useful for covert, and is certainly so for rough spots in the pleasure ground and in semi-wild places.

Erianthus Ravennæ.—This is noticed more in consequence of its being recommended in some of our seed catalogues of late than from any merit it is likely to possess for the English cultivator. Around Paris it makes a tolerably strong growth, but I fear it is not worthy of extensive cultivation there or in England.

The Ferulas.—I wish it were not necessary to write in praise of such very fine plants as these, so noble in aspect and beautiful in leaf. If you grow 2000 kinds of herbaceous plants, the first things that show clearly above the ground in the very dawn of spring (even in January) are the deep green and most elegant leaves of these ferulas. In good garden soil they look like masses of

Leptopteris superba, that most exquisite of ferns. Their chief charm will probably be found to consist in their furnishing masses of the freshest green and highest grace in early spring. The leaf is apt to lose some of its beauty and fade away early in autumn, but this may to some extent be retarded by cutting out the flower-bearing shoots the moment they appear. Not that these are ugly; for, on the contrary, the plants are fine and striking when in flower. It is indispensable that the Ferulas, like some other hardy foliaged plants, be planted permanently and well at first, as it is only when they are thoroughly established that you get their full effect. At a first view, the best way to treat them would appear to be, so to arrange them that they would be succeeded by things that flower in autumn, and only begin their rich growth in early summer; but equally wise will it be to plant them near the margin of a shrubbery, where it is desired to have a diversified and bold type of vegetation. In the rougher and more solid ground, so to speak, near large rockwork or rootwork, they would of course prove grand. The Heracleums, so often recommended in garden literature for planting near water, &c., are mere coarse rags compared to the Ferulas, while the Ferulas may be used in the places recommended for Heracleums. We may look forward to a day when a far greater variety of form will be seen in English gardens than is at present the case, and these Ferulas are thoroughly well worth growing for their superb spring and early summer effect. The best species are F. communis, tingitana, and neapolitana. Probably a few others, including sulcata, ferulago, and glauca, may with advantage be added where variety is sought, but the effect of any of the first three cannot be surpassed. Among "aspects of vegetation" which we may enjoy in these cold climes, nothing equals that of their grand leaves, pushing up with the snowdrop. In semi-wild spots, where spring flowers abound, it will prove a most tasteful and satisfactory plan to drop a Ferula here and there, in a sunny spot, and leave it to nature, and its own good constitution ever afterwards.

Gynerium argenteum (*the Pampas Grass*).—This is so well.

known to the reader that there is no excuse for naming it here except
an opportunity to say a few words as to the splendid use we may
make of it in the branch of gardening we are now discussing. It

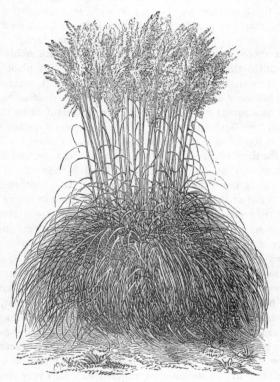

FIG. 15.—Gynerium argenteum.

is to the Dublin Botanic Gardens we owe the introduction of this
noble plant, now grown in every country where ornamental gardening is pursued. It really deserves as much attention as any

plant in cultivation, and yet how rarely is any thorough preparation made for its perfect development. A paltry class of tender plants may cost more labour and time in the course of a few months than would suffice to plant a field of the Pampas grass, yet such a glorious thing as this may be put in with perhaps a barrowful of mould to start on a bad soil, and then perhaps be placed by the water or some other secondary spot called its "proper place." What is there growing in garden or in wild more nobly distinct and beautiful than the great silvery plumes of this plant waving in the autumnal gusts—the burial plumes as it were of our summer too early dead ? What tender plant so effective as this in giving a new aspect of vegetation to our gardens if it be tastefully placed and well-grown ?   Long before it flowers it possesses more merit for its foliage and habit than scores of things cultivated indoors for their effect—Dasylirions, &c., for example, and it would be well worthy of being extensively used if one of its silken-crested wands never put forth in autumn.   It is not enough to place it in out-of-the-way spots—the general scene of every garden and pleasure ground should be influenced by it—it should be planted even far more extensively than it is at present, and given very deep and good soil either natural or made.   The soils of very many gardens are insufficient to give it the highest degree of strength and vigour, and no plant better repays for a thorough preparation, which ought to be the more freely given when it is considered that the one preparation suffices for many years.   If convenient, give it a somewhat sheltered position in the flower garden, so as to prevent as much as possible that ceaseless searing away of the foliage which occurs wherever the plant is much exposed to the breeze.   We rarely see such fine specimens as in quiet nooks where it is pretty well sheltered by the surrounding vegetation.   It is very striking to come upon noble specimens in such quiet green nooks; but, as before hinted, to leave such a magnificent individual out of the flower garden proper is a decided mistake.

HELIANTHUS ORGYALIS.—They use this in some parts of the

Continent as an ornamental-leaved plant in the pleasure ground, &c. It is as hardy as the common dandelion, grows to a considerable height, and is of a very distinct habit. Its distinction arises from the fact that the leaves are recurved in a peculiarly graceful manner. At the top of the shoots indeed their aspect is most striking, from springing up in great profusion and then bending gracefully down. It will form a capital subject for the group of fine-leaved, hardy plants, not running through the ground and requiring all the room for itself to spread about. As it is apt to come up rather thickly the cultivator will act judiciously by thinning out the shoots when very young, so that those which remain may prove the stronger and the better furnished with leaves.

Hibiscus roseus.—This is a very noble hardy perennial, growing from four to six feet high about Paris, and having the upper part of each of its abundant shoots set thickly with buds which produce flowers fully six inches across, of a showy rose, with straight deeply coloured veins running from the rich dark crimson base of the petals, and gradually becoming lost towards the margin. There is reason to think it thoroughly hardy, and it is well worth a trial in good soil in the southern and milder parts of England and Ireland. The show it makes in autumn is really very fine, and it will probably be found a grand thing for association with noble autumn flowers, like the Tritoma and Pampas grass. As regards leaf effect, it is scarcely subtropical—to use again that awkward term—and should perhaps be classed with showy herbaceous plants; but as it was used with pretty good effect in one of the Paris parks, I name it here. It should have a warm position, and deep, rich, and light soil.

Macleaya cordata.—This is a fine plant in free soil, but comparatively poor in that which is bad or very stiff. It is quite distinct in habit and tone, and sometimes goes beyond six feet high. The flowers are not in themselves pretty, but the inflorescence when the plant is well grown has a distinct and pleasing appearance. It will prove a good thing for associating with other fine hardy plants suitable for making bold groups. With some of the things before named, and with other perfectly hardy plants, there should be no difficulty in producing as bold and striking groups of vegetation as any ever seen either with us or in Paris, and afforded by costly and tender exotics requiring winter protection.

MELIANTHUS MAJOR.—This is usually treated as a greenhouse plant, and is sometimes put out of doors in summer. So treated, however, the full beauty of the plant has not time to develope; and

FIG. 16.—Melianthus major.

much the better way is to treat it as a hardy subject, putting it out in some sunny and sheltered spot, where the roots will not suffer from wet in winter. The shoots will be cut down with frost, but the root will live and push up strong ones in spring, forming by midsummer a bush of very distinct and beautiful leaves.

I have grown it in this way to a much more presentable condition than it ever assumes indoors, where it is usually drawn too much. I used to protect the roots in winter by placing leaves over them, and then covering all with a handlight, but have seen the plant survive without this protection. It is, however, best to make quite sure by using protection, except where the soil and climate are particularly favourable.

MOLOPOSPERMUM CICUTARIUM.—There is a deep-green and fern-like beauty displayed profusely by some of the umbelliferous family, but I have rarely met with one so remarkably attractive as this species. Many of the class, while very elegant perish quickly, get shabby indeed by the end of June, and are therefore out of place in the tasteful flower garden ; but this is firm in character, of a dark rich green, stout yet spreading in habit, growing more than a yard high, and making altogether a most pleasing bush. It is perfectly hardy, a native of Carniola, easily increased by seed or division, but very rare just now.. I doubt if it is even in our botanic gardens, but hope to see it in cultivation ere long.

NICOTIANA MACROPHYLLA (Fig. 17).—This is simply a garden name for a fine large variety of the common tobacco. As it is so readily raised from seed, and grows luxuriantly in rich soil, I need not say it is a very desirable subject for association with the castor-oil plant and the like, and especially suited for the many who desire plants of noble habit, but who cannot preserve the tender ones through the winter under glass. The flowers are very ornamental. It should be raised on a hotbed, and put out in May.

PANICUM BULBOSUM is a tall and strong grass, with a free and beautiful inflorescence. It grows about five feet high, and the flowers are very gracefully spread forth. It forms an elegant plant for the flower garden, in which grace and variety are sought; for dotting about here and there, near the margins of shrubberies, &c. ; and indeed for the sake of its flowers alone. P. virgatum is also a good bold grass. Both of these may be raised from seed, and are well worthy of cultivation.

FIG. 17.—Nicotiana macrophylla.

PHYTOLACCA DECANDRA.—The true plant of this name forms a very free and vigorous mass of vegetation, and though perhaps scarcely refined enough in leaf to justify its being recommended for flower garden use, no plant is more worthy of a place wherever a rich herbaceous vegetation is desired : whether near the rougher approaches of a hardy fernery, open glades near woodland walks, or any like positions.

POLYGONUM CUSPIDATUM.—This is an unusually large herbaceous species of a genus which, as cultivated in our botanical collections, does not appear likely to afford an elegant or a graceful subject for our gardens. But it is one of the best hardy things which can be recommended for their embellishment. The growth is rapid, the size unusual, perhaps eight or ten feet in very good soil, and the bearing of the plant not at any season shabby. It is covered with flowers in autumn. The same plant is often called P. Sieboldi, and frequently sold by that name. When planted singly, and away from other subjects, its head assumes a rather peculiar and pretty arching character, and therefore it is not quite fit for forming centres or using in groups, so much as for planting singly on the turf, there leaving it to take care of itself and come up year after year. In this way it would be particularly useful in the pleasure ground or diversified English flower garden. It is also good for any position in which a bold and distinct type of vegetation is desired, while of course, when we come to have fine groups of hardy " foliage plants " in our gardens, its use will be much extended. The deeper and better the soil, the finer will its development prove. You cannot make the soil too deep and good if you want the plant to assume a fine character. As with tender plants we have no end of attention to bestow, often daily attention, the time and labour necessary to well prepare the ground for a hardy subject should never be begrudged. This plant will probably be also found useful for game covert. It is easily procured in our nurseries, and there is plenty of it at Kew, or used to be.

RHUS GLABRA LACINIATA.—We have known this plant for about two years as a subject of much promise for garden decoration, and may now be certain of its being one of the most useful and elegant dwarf shrubs we can employ to furnish an attractive effect from leaves. It is a small kind, with finely cut and elegant leaves, the strongest being about a foot long when the plants are established a year or two. When seen on a nicely established plant, these leaves combine the beauty of those of the finest Grevillea, with a fern frond, while the youngest and unfolding leaves remind one of the dainty ones of a finely-cut umbelliferous plant in spring. The variety observable in the shape, size, and aspect of the foliage makes the plant charming to look upon, while the midribs of the fully grown leaves are red, and in autumn the whole glow off into bright colour after the fashion of American shrubs and trees. During the entire season it is presentable, and there is no fear of any vicissitude of weather injuring it. Its great merit is that in addition to being so elegant in foliation it has a very dwarf habit, and is thoroughly hardy. Plants at three years old and undisturbed for the past two years are not more than eighteen inches high. The heads are slightly branched, but are not a whit less elegant than when in a simple-stemmed and young state, so that here we have clearly a subject that will afford a charming fern-like effect in the full sun, and add graceful verdure and distinction to the flower garden. When the flowers show after the plant is a few years old they may be pinched off, and this need only be mentioned in the case of permanent groups or plantings of it. To produce the effect of a Grevillea or fern on a small scale, we should of course keep this graceful Rhus small, and propagate it like a bedding plant. The graceful mixtures and bouquet-like beds that might be made with the aid of such plants need not be suggested here, while of course an established plant, or groups of three, might well form the centre of a bed. Planting a very small bed or group separately in the flower garden and many other uses, which cannot be enumerated here, will occur to those who have once tried it. Some hardy plants of fine foliage are either so rampant or so top-heavy that they cannot be wisely associated with bedding plants—this is, on the contrary, as tidy and tractable a grower as the most fastidious could desire. It would be a pity to put such a pretty plant under or near rough trees and shrubs—give it the full sun, and good free soil.

SEDUM SPECTABILE.—This is one of the finest autumn flowering plants introduced of late years, being at once distinct, perfectly hardy, fine in flower, and pretty before it unfolds from its dense bush of glaucous leaves. It is hardly large enough to be included here, but is so valuable for association with the nobler hardy plants in beds, for use around shrubberies, as a pot plant, a rock plant, or a first-class border plant, that I cannot pass it by. When specimens of it are fully exposed to the sun and air, and well-established—which they become in a year or so—it is particularly fine, and flowers till the season is just over, keeping company with the Tritomas. It begins to push up its fat glaucous shoots in the very dawn of spring, keeps growing on all through the early summer, and continues in a perfectly presentable condition. The plant is known by the name of "fabarium" or "fabaria" in our gardens, and that has caused some little confusion, as the true S. fabaria of Koch is quite distinct from, and a very poor plant compared with it. It was in the first place named fabarium by M. C. Lemaire, but as this was likely to cause some confusion, M. Boreau named it spectabile, which, considering its very noble character as a Sedum, and the desirability of having a distinct name for a plant so distinct, is on the whole the best. The plant is one of the easiest to propagate and grow that has been introduced to this country, and deserves to rank among the very best herbaceous subjects. It is extensively employed in Parisian gardens.

THE TRITOMAS.—So hardy, so magnificent in colouring, and so fine and pointed in form are these plants, that we can no more dispense with their use in the garden where beauty of form as well as beauty of colour is to prevail, than we can with the noble Pampas grass. They are more conspicuously beautiful when other things begin to succumb before the gusts and heavy rains of autumn, than any plants which flower in the bright days of midsummer. It is not alone as component parts of large back ribbons and in such positions that these grand plants are useful, but in almost any position in the garden. Springing up as a bold close group on the green turf and away from brilliant surroundings, they are more effective than when associated with bedding plants; and

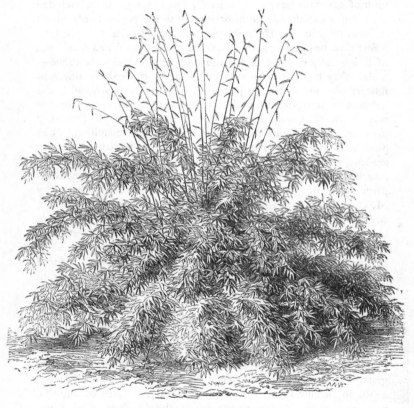

FIG. 18.—Bambusa falcata.

of course many such spots may be found for them near the **margins** of the choice shrubbery in the generality of pleasure grounds.   It is as an isolated group flaming up amid the verdure of trees and shrubs and grass that their dignified aspect and brilliant colour are seen to best advantage. However tastefully disposed in the flower garden they will prove generally useful, and particularly for association with the finer autumn-flowering herbaceous plants. It seems we do not sufficiently appreciate the advantage of good hardy plants, however much we may grumble at the consumption of coals. Here are the finest of all autumnal flowers, never causing a farthing of expense for wintering, storing, replanting, &c., but merely asking for a little ordinary preparation of the soil at first, and yet they are merely grown as adjuncts even in good gardens, and in many you can scarcely find them. For every quality that should make a plant valuable in the eyes of the flower gardener, they cannot be surpassed by any subjects that require expensive care all through the winter; indeed we may say they cannot be equalled by any of such—a sufficient proof that it is not only those who possess stoves, greenhouses, and glass-gardens, so to speak, that may enjoy the highest beauty in their gardens; and that it is not solely among tender plants we must look for subjects wherewith to carry out that most desirable end—the adding of a greater degree of interest and beauty to British flower gardens. There can be little doubt but T. glaucescens is the most generally useful kind, flowering profusely, no matter what the weather, in August and September, and coming in at a time when people frequent the country garden so much. Next to it in importance, and greater than it in stature, is T. grandis, which comes in flower very late. The stems grow to six or even seven feet high, sometimes throw out a side spike, and flower away if the season prove mild till Christmas, or even till the end of January. T. Rooperi, with which this has been confounded, appears to be quite a different plant, one of very strong agave-like habit of leaf, and flowering very late—so late indeed that it is all but useless for open-air work, though it may make a useful pot plant, and flower indoors in winter. T. serotina would seem to be a variety of T. uvaria, and not far removed from what we call grandis. T. media is not worth growing, so far as I have observed, in consequence of sending up its stems so very late in autumn. The best of all those dubious kinds is one called grandiflora, which is more distinct than the others, forming a sort of large corm-like base, and producing, when in good soil, large heads of finely coloured flowers. So much for these not often seen kinds or species. The important fact is that we have two kinds, glaucescens and grandis, that make a splendid display on any soil, and only require tasteful planting and arrangement to produce a noble feature in the flower garden in autumn.

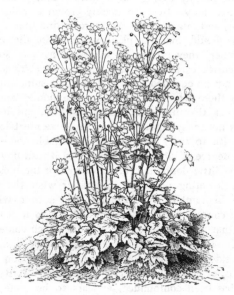

FIG. 19.—Anemone japonica alba.

The deep London clay is highly inimical to most kinds of herbaceous plants, but by making some preparation in it for the Tritomas I have found them do nobly, and they may be grown to perfection in all parts of these islands without any trouble beyond planting in good and deep soil, and with some deepening and " making " of the soil in poor and very shallow ones.

A most satisfactory result may be produced by associating these Tritomas with the Pampas and the two Arundos, the large Statice latifolia and the strong and beautiful autumn flowering Anemone japonica alba (See Fig. 19). This is peculiarly suited for association with hardy herbaceous plants of fine habit, and should be in every garden where a hardy flower is valued.

VERBASCUM VERNALE.—Most of us know how very distinct and imposing are the larger verbascums, and those who have attempted their culture must soon have found out what transient far-seeding things they are. Of a biennial character, their culture is most unsatisfactory: they either migrate into the adjoining shrubbery or disappear altogether. The possession of a thoroughly noble perennial one must therefore be a desideratum, and such a plant will be found in the Hungarian Verbascum vernale. This is fine in leaf and stature, and produces abundance of flowers. The lower leaves grow eighteen or twenty inches long, and the plant when in flower to a height of seven or eight feet, or even more when in good soil. It is a truly distinct subject for helping us to vary matters, and may, it is to be hoped, ere long be to be had in our gardens and nurseries. It is a scarce plant in England, and perhaps not as yet to be had in many of our nurseries or botanic gardens, though it is certainly the best plant of the genus as known to us in gardens. I first saw it in the Jardin des Plantes.

THE YUCCAS.—Among all the hardy plants ever introduced to this country, none surpass for our present purpose the various kinds of Yucca, or "Adam's Needle," as it is sometimes called. There are several species hardy and well suited for flower garden purposes, and, more advantageous still, distinct from each other. The effect afforded by them, when well developed, is equal to that of any hothouse plant that we can venture in the open air for the

summer, while they are hardy and presentable at all seasons. They
may be used in geometrical " English," or any other style of garden ;
may be grouped together on rustic mounds, or in any other way the
taste of the planter may direct. The best perhaps, considering its
graceful and noble habit, is Y. recurva, which is simply invaluable
in every garden. Old and well established plants of it standing
alone on the grass are pictures of grace and symmetry, from the
lower leaves which sweep the ground to the central ones that point
up as straight as a needle. It is amusing to think of people putting
tender plants in the open air, and running with sheets to protect
them from the cold and rain of early summer and autumn, while
perhaps not a good specimen of this fine thing is to be seen in the
place. Than this there is no plant more suited for planting between
and associating with flower beds. Next we have Y. gloriosa, more
pointed in habit and rigid in style, and also large and imposing in
proportions. Lacking the grace of recurva, it makes up for that to
some extent by boldness of effect, while, like the preceding, it sometimes sends up a huge mass of flower. Y. gloriosa varies very much
from seed—an additional recommendation, as the more variety of
fine form we have the better. Then there is Y. glaucescens, with
a sea-green foliage, and rather free to flower, the buds being of a
pink tinge, which tends to give the whole inflorescence a peculiarly
pleasing tone. This is a first-class plant. Y. filamentosa is smaller
than these, but one which flowers with much vigour and beauty.
It is well worth cultivating in every garden ; not only in the flower
garden or pleasure ground, but also for the rough rockwork, or any
spot requiring a distinct type of hardy vegetation. Yucca flaccida
is somewhat in the way of this, but smaller. It flowers even
more abundantly and regularly than filamentosa, and is well worthy
of cultivation. The preceding species, if not so much used in our
gardens as they deserve, are at all events known in them. The
following I met with for the first time in Parisian gardens:
Y. lutescens. This is a species of neat habit and slightly yellowish
tone, of shining green, and very distinct. Y. flexilis is an ornamental, though not large growing kind. Y. stricta is a rigid species

scarcely so effective as the preceding; and Y. angustifolia has narrow pointed leaves and a distinct habit. Y. Treculeana is a very noble species, which will be found perfectly hardy on good soil and in warm situations. It has deeply furrowed and very large rigid leaves, and is well worthy of culture even in a cool house, in which it is sometimes kept in this country. If we had but this family alone, our efforts to produce an agreeable effect with hardy plants need not be fruitless. The freely flowering kinds, filamentosa and flaccida, may be associated with any of our nobler autumn flowering plants, from the Gladiolus to the great Statice latifolia—the species that do not flower so often, like recurva and gloriosa, are simply magnificent as regards their effect when grown in the full sun and planted in good soil; and I need not say bold and handsome groups may be formed from devoting isolated beds to this family alone. They are mostly easy to increase by division of the stem and rhizome; and should in all cases be planted well and singly, beginning with healthy young plants, so as to secure a perfectly developed single-stemmed specimen.

Hardy Fern-like Plants for the Flower Garden.

However much we may appreciate the grace of ferns in the conservatory or plant-house, we have yet but very rarely employed them in the flower garden, and there are obstacles which are likely to prevent their use in such; but it has often struck me that we may amongst herbaceous plants find many things that afford all the grace of a fern, and yet withstand the sun as well as a stonecrop. I have frequently drawn attention to several good things in this way, and particularly to Thalictrum minus, which, while growing freely in ordinary soil in the full sun, and being perfectly hardy and permanent, will afford us a beauty almost identical with that of Adiantum cuneatum, and which may be made the nicest use of by the flower gardener. In consequence of recommending it so strongly through various channels, there have been many demands for the plant. It is to be had in abun-

dance in some parts of Britain, particularly in the north and
north-west.　It is also freely found in Ireland, and is abundant in
the Lake District, growing high up amongst T. alpinum, and a
taller and coarser species.　In the Jardin des Plantes, I ascertained
that T. fœtidum, a dwarf, slightly glaucous species, with very
elegantly divided leaves, is much better than even the best form of
T. minus.　Once established, the only thing that need be done is
to pinch off the growing flower stems, and thus keep the poor
little flowers out of sight.　The Italian Isopyrum thalictroides is
dwarfer in growth and with a similar aspect, though the leaflets are
larger.　It is grown in most botanic gardens, and may doubtless be
had in profusion in its native country.　It would form a graceful
dwarf fern-like fringe, but is not equal to the pretty T. fœtidum
when once firmly established in nice cushiony tufts.　It is doubtful
if we shall ever surpass or equal that as a little fern-like plant for
the flower garden.　The Isopyrum must also have its young flower
shoots pinched carefully off.　There is another fern-like plant useful in the same way, quite distinct in aspect from either of the foregoing, and likely to furnish a most useful fern-like effect.　It is Spiræa
filipendula, the dropwort—either the single or the double kind will
do, the last the best.　The plant is found frequently wild in England.
The leaves are cut into deeply toothed segments, will of course
stand any amount of exposure, are pleasing in outline, lasting in
character, and certain to produce a good fern-like effect of the
pinnated type.　Pinch off the stems and you will then have no
further trouble in producing a dense green margin with this plant.
The leaves will grow from five to eight inches long, according to
the soil.　The flower gardener who is at all inclined towards variety
and interest, can of course make a tasteful use of these plants.
The leaflets of T. fœtidum may be used among flowers with good
effect.　They are of a more lasting character than those of the
maiden-hair fern.

ANTHRISCUS FUMARIOIDES is a lovely fern-like plant, as hardy
as the common parsnip, a native of Croatia, and fresh as an emerald

to the very end of autumn. It is one of those plants occasionally
to be found in botanic gardens, but which the general public cannot
readily get. Ever since I first became acquainted with the dark green
and delicately graceful Meum athamanticum, I have had the fullest
confidence that from this Parsley order we shall yet get some of the
most charming ornaments for our flower gardens. However, in
looking through the species with a view of selecting some for this
purpose, we find that, however beautiful are the leaves of some
umbellates, their value is but slight in consequence of their perishing
early in the year. I made a point of looking at this family every
time I went to the Jardin, and found that the following species
with elegant leaves preserved their verdure sufficiently long to
make them valuable as flower garden or pleasure ground ornaments; indeed, some of them were as green as an emerald at the
end of the season : Seseli elatum, globiferum, and gracile, Atha-
manta Matthioli, Silaus tenuifolius, Meum adonidifolium, Peuce-
danum longifolium, Petteri, and involucratum, and Anthriscus fumarioides. These were all in prime verdure the last week in August,
and would no doubt preserve their freshness to the last moment in
these islands. To these I would also add, as " fern-leaved plants,"
Pyrethrum tanacetiodes and achillæfolium, and Tanacætum elegans,
a distinct silvery-leaved species.

List of Hardy Herbaceous and Annual Plants, &c., of fine habit, worthy of employment in the flower garden or pleasure ground.

Acanthus, several species.
Asclepias syriaca.
Statice latifolia.
Morina longifolia.
Polygonum cuspidatum.
Rheum Emodi, and several other species.
Euphorbia Cyparissias.
Datisca cannabina.
Veratrum album.
Tritomas, in variety.
Thalictrum fœtidum.
Crambe cordifolia.
Althæa taurinensis.
Geranium anemonæfolium.
Melianthus major.
Panicum, several species.
Spiræa Aruncus.
,,   venusta.
Astilbe rivularis.
,,   rubra.
Eryngium, several species.
Ferula,   ,,
Seseli,   ,,
Chamærops excelsa.
Cucumis perennis.
Hibiscus roseus.
Rhus glabra laciniata.
Artemesia annua.
Phytolacca decandra.
Centaurea babylonica.
Lobelia Tupa.
Pencedanum ruthenicum.
Heracleum, several species.

Dipsacus laciniatus.
Alfredia cernua.
Cynara horrida.
,,   scolynus.
Carlina acanthifolia.
Telekia cordifolia.
Echinops exaltatus.
,,   ruthenicus.
Helianthus argophyllus.
,,   orgyalis, and others.
Gunnera scabra.
Funkia subcordata.
,,   japonica.
Tritoma, in varieties.
Arundo Donax.
,,   conspicua.
Gynerium argenteum.
Elymus arenarius.
Bambusa, several species.
Arundinaria falcata.
Yucca, several species.
Verbascum vernale.
Aralia spinosa.
,,   japonica.
,,   edulis.
Macleayia cordata.
Panicum bulbosum.
,,   virgatum.
Kochia scoparia.
Datura ceratocaula.
Silybum eburneum.
,,   marianum.
Onopordon Acanthium.
,,   arabicum.

List of Hardy Plants of fine habit, that may be raised from Seed.

Among suitable hardy plants that may be raised from seed, the following are offered in the seed catalogues for the present year :—

Acanthus latifolius.
 „ mollis.
 „ spinosus.
Artemesia annua.
Astilbe rivularis.
Campanula pyramidalis.
Cannabis gigantea.
Carlina acanthifolia.
Datura ceratocaula.
Echinops, several species.
Eryngium bromeliæfolium.
 „ campestre.
 „ cœlestinum.
 „ giganteum.
Ferula communis.
 „ tingitana.
Geranium anemonæfolium.
Gunnera scabra.
Gynerium argenteum.

Helianthus argyrophyllus.
 „ orgyalis.
Heracleum eminens.
 „ giganteum.
 „ platytænium.
Kochia scoparia.
Lobelia Tupa.
Morina longifolia.
Onopordon arabicum.
 „ tauricum.
Centaurea babylonica.
Panicum, several species.
Phytolacca decandra.
Salvia argentea.
Silybum marianum.
 „ eburneum.
Statice latifolia.
Tritomas, in variety.
Yucca, several species.

CHAPTER III.

Le Jardin Fleuriste de la Ville de Paris:

THIS, the great nursery and propagating garden for the tender plants employed for the decoration of such of the parks and gardens as belong to the city, and also for decorating the Hôtel de Ville on festive occasions, is so very remarkable an establishment, that no excuse is needed for publishing a description of it. In Paris of recent years the making of new gardens has had as much attention as the pulling down of old streets, and the richness and extent of the collections in this establishment are quite astonishing even to visitors from our own great gardens. In addition to La Muette—the name by which the establishment at Passy, that I am about to describe, is generally known—there are two other establishments for the supply of the city with plants —one for trees and shrubs in the Bois de Boulogne, and the other for herbaceous plants, &c., in the Bois de Vincennes. It should also be remarked that it is only for the supply of the gardens of the *Ville*—as distinguished from those of the State. The gardens of the Tuileries, Luxembourg, at the Museum, &c., grow their own supplies.

Imagine yourself prepared to visit a " propagating establishment," and then ushered into a magnificent span-roofed curvilinear camellia house—quite a grand conservatory of camellias—and in connexion with it on one side a great conservatory filled with Aralias, Yuccas, Beaucarneas, tree ferns, Nicotianas, Dasylirions, Dracænas, and a host of such plants all in fine condition, and arranged as closely and well as possible, everything being clean and orderly. Then, on the other side, another very fine span-roofed structure for palms. And such a noble collection of healthy palms in a fresh green state! They were arranged in three longitudinal beds, while all along the sides of the house ran a belt of the smaller and younger kinds, plunged in tan to give them a little encouragement. To look along the pathway between these long beds was like glancing into a fresh young tropical palm grove, in such perfect health were the plants. When it is considered that many other great houses are in the garden, and a large field of pits and frames, the reader will agree that examining or relating the particular interest of each subject is out of the question, and particularly when it is stated that there are nearly 400 kinds of palms alone in this establishment. Though it is essentially a business garden, and one in which an almost innumerable host of plants have to be annually developed, no slovenliness of arrangement or culture was apparent in any part. The plants generally are clean, well-grown, and well-arranged. Seldom indeed do we see such efficient economy of space in gardens as is the rule in these houses. Under the benches are packed quantities of Caladiums, Fuchsias, Cannas, and hosts of things that may be efficiently preserved in such places in winter; and even after the great Arums, &c., are potted off in spring, they are thickly placed underneath for a short time, every available inch being taken advantage of. Some of the houses are large lean-to's, and instead of the back wall being left naked, or with one shelf placed against it at the top, there are a series of shelves one above another, six altogether, and on these a multitude of plants are accommodated—Coleuses, &c., in the warm houses; Lantanas, and the like, in the cool; they keep readily enough on these during the winter, and, if drawn a little or discoloured, a few weeks of bright sun in spring in the frames before putting out in the open air, soon puts that to rights. In the large span-roofed curvilinear houses, with a narrow passage through the centre. there are a series of shelves affixed to erect irons on each side of the

F

central pathway, and on these a great number of plants are stored, so that every space is taken advantage of without in the slightest degree interfering with the health of the plants, which is indeed admirable. But doubtless it is necessary thus to economize space, for the enormous number of nearly three million of plants is annually required from and furnished by this establishment for the embellishment of Paris and its environs. These are raised at a very cheap rate—less than a penny each. And observe that many of the plants are such as would be fit to embellish any exhibition, numbers of them being palms and fine-leaved plants, while of course the least valuable are simply bedding plants, from Nierembergias to Pelargoniums, of which last there are annually sent out from hence 400,000 plants !

Without seeing the houses or plants, the potting-shed would tell of extraordinary horticultural operations, for in the centre there is a great wide bench, around which sixty men can work. Ordinary bedding plants are kept here at an extraordinarily cheap rate—I think more economically than I have observed them in English gardens. A very large space of ground is covered by parallel lines of rough and rather shallow small wooden frames, very simply and cheaply made. They are rather closely placed, and the pathways between, and, in fact, all the spaces around them, filled up with leaves and mossy rakings from the adjacent Bois de Boulogne. These are nearly or quite filled up to the edge of the frames, and of course keep the plants warm and snug through the winter. In winter the floor of the frames is low; in spring their bottoms are raised so as to bring the foliage of the plants right up to the glass; and the men were engaged in doing this by putting in a quantity of the pretty well consolidated leafy stuff before named. When this becomes decomposed around the rough and simple frames, it is taken away and preserved for potting purposes, making of course excellent leaf-mould. With the frames thus plunged in a comforting mass of genially warm leaves, and the plants lowered inside in winter, protection is a matter of easy accomplishment. They were crammed with healthy bedding plants—the most precious of the variegated Pelargoniums, &c., being of course in the houses; but the great mass of bedding stuff is grown in this way in miles of cheap framing. There is no means of heating these otherwise than by the leaves; and as it would in any case be very desirable to accumulate a goodly store of leaf-mould, that means is anything but expensive.

The foregoing is a very common method for protecting plants in France, and this description of it was written early in the spring of 1867. But things are only beginning at La Meutte, from what we hear of future glasshouse buildings. The municipality have arranged to build sixty more of them in this really excellent town-garden. Apart altogether from this series, a number of houses were erected at La Muette during the past summer, which materially encroach upon the space occupied by the rough framing alluded to. These houses are especially intended for bedding plants, and are so well adapted for that end that some details about them may be useful. They have been designed on an excellent plan for the culture of bedding plants, raising of seedlings, &c., growth of seedling palms, and in a word, of all dwarf plants. I have seen a good many houses devoted to similar purposes in all parts of these islands, but never in public, private, or commercial garden anything so complete in its way as the block of houses, of which Fig. 20 represents part of the end view. They are low, and rather narrow, so that all operations may be conducted from the central pathway.

They are cheaply made of thin iron, and the roof consists of one sash at each side. Many of

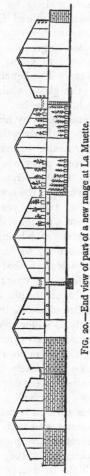

FIG. 20.—End view of part of a new range at La Muette.

F 2

the frames before alluded to were furnished with iron sashes; and many of these were utilized in the building of the houses. Passing along by the ends of these houses you may see a bench about a hundred feet long, filled completely with the deeply-dyed Alternantheras—a sheet of colour; the next devoted to young palms, as green and vigorous as if in their native wilds; another devoted to young Dracænas and fine-leaved plants generally; and so on. The benches are of slate, and the plants are held well up to the glass, while quantities of plants in the way of Canna and Dahlia may be stored beneath. We generally prefer wooden houses, but any horticulturist who has seen the plants in the low range at Passy will agree with me that no plants were ever seen in finer health or condition than the numerous species in it; while the very permanent nature of the structure is a great gain, inasmuch as a wooden series of the same character would require a complete overhaul in the course of a dozen, and perhaps reconstruction at the end of twenty years. A mode of protecting these houses in very severe weather is deserving of notice. It is by means of wooden shutters, each being about the size of the sash of the house. As will be seen by the engraving, the gutters, strongly lined with zinc, are wide, so that men can run along with the greatest ease to perform any operation between them. In winter, the gutter space is frequently filled with leaves, to prevent the influence of frost settling down between each house; while the glass of the houses likely to suffer most is protected at night by the shutters. These are not taken from between the houses every day, but simply left in piles of ten or so over some unoccupied spot, or if the range happens to be completely filled, shifting the position devoted to each pile of shutters every day, so as to prevent the plants beneath from suffering. The facility and simplicity with which these houses may be thus encased in wood to meet a very severe frost, and in a few minutes, and without the least untidiness of any kind, during day or night, are quite admirable. Matters are so arranged in the houses that they could dispense entirely with this precaution, which is simply noticed from its adaptability to many places

where a great number of bedding plants have to be kept,
and where means of heating sufficiently to keep out very severe
frosts, are not forthcoming. But the ground plan of the range
is also worthy of particular notice. By thoughtful attention to
that, the men at work in any one of the eighteen houses of
the block already completed may pass and convey plants from
one to the other without passing through the open air; and thus
one important point, both as regards the comfort of the men and
the health of the plant is secured, and in this way. To put it
simply, the great mass of long houses is cut in two by a covered
glass passage, or rather all the houses communicate with it. This
simple diagram will show the arrangement at a glance, and the

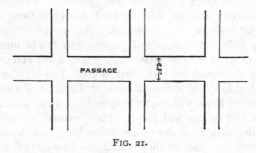

PASSAGE   6.FT.

FIG. 21.

passages of the various houses diverging from the central covered
way. Already nine houses are arranged on each side of this
passage, and it is proposed to continue the arrangement till all the
ground previously devoted to framing is covered with this excellent
class of house. The visitor, entering at the end shown by Fig. 20,
and continuing his way through the first house, would at its
further end meet with the covered way running at right angles to
the houses, and through this he could enter any of the other houses
he wished to see without again exposing himself or opening any
doors to chill the plants in winter, or running the draughty gauntlet,
which he usually has to run where houses are arranged in the
ordinary scattered way. Moreover, as in many cases one long

house is devoted to a particular species or variety in much request, the visitor may see the state of the stock by simply traversing the passage, and looking through the glass dividing it from the houses.

But though the ordinary dwarf bedding plants are preserved in vast quantities both in the frames and these houses, it is not the cheapest way in which they do things here, as we shall presently see. You have heard of the grand and graceful use made of the Cannas in Parisian gardening. These are preserved in a most efficient way in caves under the garden. When the stone is taken out of the ground for building purposes, an odd rough propping column is left here and there, and thus wide and spacious cellars of equable temperature are thus left underground. They are in this case about seven feet high, and are used as a great store for plants that may be well preserved without light in the winter. You descend by a sloping tan-covered passage, and ten to one but you imagine yourself in a large potato store immediately you get down, even if you are acquainted with horticulture. I did, and was it any wonder, when heaps of different kinds of Canna, and those by no means common kinds with us, were spread upon the floor a yard or more deep, and twenty feet long? The "tubers" of some of the large varieties were from five to ten inches long, and the men were turning them over just as they would the contents of a series of potato-pits. Here too in wide masses against the wall are arrayed quantities of Aralia papyrifera, the handsome and much grown species so useful for "subtropical gardening." It seems in perfectly firm and safe condition, growing in this dark, or rather gas-lighted climate, and sends out long blanched leaves of a delicate lemon colour, which will of course soon acquire a healthy green when the plants are placed in the open air. Thus they preserve Aralia papyrifera in all sizes, and this fine thing is turned out for garden embellishment almost as cheap as wallflowers. Of course analogous protection could be given to such things in many English gardens where space may be limited, and much expense out of the question. In these caves were also preserved Brugmansias,

American and other Agaves, Dahlias, Fuchsias, &c., *ad lib.*, and it seemed to me about the best possible place for storing such plants. It is astonishing the quantities in which you see rare things and new bedding plants here. Houses, eighty and one hundred feet long, are filled with one variety; houses equally long, devoted to the raising of seedling palms, &c., in quantity. If a plant be considered worthy of attention at all it is propagated by the thousand; 30,000 being the opening quantity for a new thing of any promise. During the past autumn 50,000 cuttings of one kind of Fuchsia were inserted here in one week. Dracænas are grown here more abundantly than variegated Pelargoniums in many a good English bedding garden, and they have, it is believed, the finest collection of them in existence. In one house a specimen of each kind has been recently planted out for trial in the central pit, and among them many handsome and noticeable kinds worthy of extensive use with us. The main entrance is in the Avenue d'Eylau, and near it there are some interesting hardy plants, including a collection of bamboos put out to test their hardiness; and that out of door matters are not forgotten will be apparent when it is stated there is a framework as big as a large conservatory, over a collection of tulips. It is a favourite plan here to devote a house to a special subject. Thus there is a large and fine span-roofed stove for Ficuses; a house for the collection of bedding Musas, with a line of thirty healthy plants of Musa Ensete, forming its backbone, so to speak; a very large and high curvilinear stove for the great collection of Solanums; special houses for Arums, Caladiums, &c., and a winter garden about 120 feet long by 40 wide, well stored with a healthy stock of usual conservatory plants, with here and there fine-leaved things like Phormium tenax, a very effective plant when well grown in pots and tubs, and of which they have here thousands of plants of various sizes. Of course all this vast collection cannot be and is not used for summer decoration. It is employed for the decoration of the Hôtel de Ville, where 10,000 plants are sometimes required upon a single occasion. Each van that conveys the plants to the Hôtel de Ville is fortified with a neat little stove at one end, flat hot-water pipes passing round the interior, and thus, while space is not curtailed, the van is efficiently heated, and tender plants can be conveyed by it in safety in the depth of winter.

Fig. 22.—Propagating-house at La Muette.

The largest series of houses grouped together for the culture of house plants generally, and the stock intermediate in size between the dwarf and bedding plants in the low houses and the tall palms and like subjects arranged in great iron conservatories, are arranged on precisely a similar ground-plan to the low range before described, but are very fine and lofty curvilinear houses. Entering from the Avenue d'Eylau, one of the first things you meet with is a magnificent block of curvilinear span-roofed houses— ten in a mass, divided by a narrow covered way which you enter, and through which only you communicate with the houses which it divides— five on each side, These contain a vast amount of interest and a great number of rare and new plants. One is the largest and most perfect propagating-house I have ever seen, being more than eighty feet long and twenty-four wide, or thereabouts. From this house immense quantities of plants are turned out in the course of a year, and observe that numbers of these are large-leaved Ficuses and plants difficult to strike, as well as Begonias, bedding plants, and free-rooting stuff. It contains three central and two side beds, the central pits well elevated, every space in active work, and the whole presenting a most imposing array of large bell-glasses.

The propagating seemed most successful. They do not do it as we do. No pans were used in this house, but very minute pots, a shade larger than a thimble, and into each a cutting is placed, the little pots placed in the tan, and covered with large circular bell-glasses, as shown by fig. 23. The greater part of the house is occupied with these, all of a size, but there are some special arrangements for propagating the more difficult subjects, and among these may be noticed what appeared to be an improvement— bell-glasses, somewhat of the ordinary

FIG. 23.

type, with an aperture in the top about two inches in diameter, into which a moisture-absorbing bit of sponge was squeezed. Nothing could be more business-like than the arrangements for propagating in this house. During winter quantities of Hibiscuses are kept in cool houses here, treated somewhat like Fuchsias—they exist without leaves or any trouble in winter. They are now being started gently in warm-houses, and prepared for planting out. The bedding-out in and around Paris is something enormous. It is not confined to grand avenues or parks, or great centres, as with us, but is dotted about wherever there is an open space; every small square has its garden, each distinct in design; and almost every round open *place* has its flower beds filled in the early part of the year with spring flowers, and in summer with bedding and subtropical plants; so they require abundant space at head-quarters. In one high house devoted to tall palms and other effective-leaved plants may be noticed a good specimen of Musa superba; they cultivate sixteen or seventeen species of this genus. In fact, the place offers more facilities for ascertaining the most valuable members of some important genera than any in existence. The boilers of some of the smaller propagating and other houses are heated by gas, and in this way a very equable temperature is preserved. It may give some approximate idea of the collection, when it is stated that there are in cultivation here between thirty and forty kinds of Aralia, thirteen of Oreopanax, thirty-six of Anthurium, fifteen of Pothos, thirty of Philodendron, nearly one hundred and twenty of Canna, eighteen of Zamia, and more than one hundred and ten of Ficus! And probably since that time the stock has much increased, for, as before remarked, the establishment is in a state of rapid growth.

CHAPTER IV.

The Public Gardens and Parks of Paris.

PARIS, if not already the brightest and most beautiful of all cities, is in a fair way to become such; and the greatest part of her beauty she owes to her gardens and trees. A city of gardens and palaces indeed; but which is the most magnificent—the view up that splendid main avenue and garden stretching from the heart of the city to the Arc de Triomphe, or that of the Tuileries or any other architectural feature of Paris? What would the magnificent new boulevards of white stone be without the softening and refreshing aid of those noble lines of well cared-for trees that everywhere rise up around the buildings, helping them somewhat as the grass does the buttercups? The makers of Paris—who deserve the thanks of the inhabitants of all the filthy cities of the world for setting such an example—answer this question for us by making parks and gardens in every direction, and by planting such quantities of trees as no British planter would believe without having walked some dozen miles in and around Paris; and by relieving in every possible direction man's work in stone with the changeful and therefore ever-pleasing beauty of vegetable life.

A great point is gained when the public gardening of a city is not quite out of the way of its more busily occupied inhabitants; when, instead of having to go a mile or two to see a public park or garden, one can scarcely go out of doors without encountering something green and pleasant to the eye; and it is so to a great extent in Paris, and will be much more so if improvements there

progress for a few years at their present rate. In Paris the public
gardens and all that relates to them are as a rule admirably managed
and kept; faults they have, to be sure, to an English eye, but on
the whole they are magnificent. And as the example shown in this
way should be followed, so far as means will permit, in every town
or city, it may be useful to walk through them with the reader,
pointing out, as far as possible in a hurried way, their most instructive
features. Any passing visitor to Paris may see the public gardening,
however difficult it may be to see the market and fruit gardens,
and therefore a description which may help him may be useful
here. To begin, we had better consider for a moment that the
gardens of Paris are under different boards of management, if we
may so speak. Thus the gardens of the Luxembourg, the Tuileries,
and other palaces, with the Jardin des Plantes, have each their own
superintendents, and have nothing to do with what we may call the
Municipality of Paris, nor are they supplied from its great nursery,
La Muette. On the other hand, the Parc Monceau, the Bois de
Boulogne, the great belt of gardens on each side of the Champs
Elysées, and all the squares, &c., belong to La Ville de Paris, and
are managed by it. In endeavouring to give the reader a concise
idea of Paris gardens, and which may act as a guide to him, I will
commence in the Place du Carrousel, the great square between the
Louvre and Tuileries, and afterwards walk into the outer gardens of

The Louvre.

The eastern part of this square is narrower, in consequence
of the projection into it of the splendid buildings of the new
Louvre, thus projected in consequence of a difference in the line of
standing of the two palaces; and in this part of the square, called
the Place Napoleon III., there are two pretty little gardens surrounded by railings with gilt spears, and they will give the visitor
an idea how the gardens of Paris are kept. They simply display
trees and evergreens, grass as green as could be desired, ivy edgings,
and they look like oases in the centre of the great palace-surrounded
square. A word or two about the grass in Paris gardens. It is

kept perfectly neat and fresh at all times, in the very heart of the city; and even through the winter it looks as green as an emerald. They give it a top-dressing of fine and well decomposed manure with leaf-mould in April, and it is thoroughly and frequently watered with the hose in summer. Doubtless they would have a garden in the centre of the Place du Carrousel also, were it not that the traffic across it from the Rue de Rivoli to the quai and the bridges on the river side, and which enters in through arches on each side, is so great. Then by passing through the great court of the Louvre and out on the eastern side the visitor may see the garden of the Louvre, which is simply a rail-surrounded space, laid out with the usual very green and well-kept grass, round-headed bushes of lilac, ivy edgings, evergreen shrubs here and there, flowers, both spring and summer, and the best, cheapest, prettiest, and most lasting edgings in use in any garden, made of cast-iron in imitation of bent sticks. Much of this garden was once covered with old buildings and streets—even the great square just spoken of was once packed with alleys, but the recent improvements of Paris have swept all those things away, and on every side the buildings stand as free as could be desired, unlike our London ones, some of which can hardly be discovered, and which when they have an enclosed space around them, it is merely a receptacle for dead cats, &c. Numerous seats are placed against the walls of the palace, and the gardens, though not large, offer a very agreeable retreat to invalids, children, and others during all but the colder months.

The main feature of the flower gardening is a modification of the mixed border system, pretty, and also capable of infinite change. This is the plan of the Louvre borders. It is a combination of circle, and mixture, and ribbon, quite unpractised with us. Along the middle of the borders we have a line of permanent and rather large-growing things—roses, dahlias, neat bushes of Althæa frutex, and small Persian lilacs. The lilacs might be thought to grow too gross for such a position, but by cutting them in to the heart as soon as they have done flowering the bedding plants start with them on equal terms, and the lilacs do not hurt them by pushing out again, and making neat round heads for taking a lead in the display of spring flowers. Thus they have along the centre of each border a line of green and pointed things, which always save it from over-colouring, and then underneath they lay on the tones as thick as need be. Around each bush or tallish plant in these borders are placed rings of bedding plants—Fuchsia, Veronica, Heliotrope, Chrysanthemum grandiflorum, fœniculaceum, &c., the outer spaces between the rings being filled with plants of other sorts. Then follows a straight line of Pelargoniums—scarlet, white, and rose mixed plant for plant, and forming a very pretty line. Outside of that a band of Irish ivy, pegged close to the earth, and pinched two or three times a year; and finally, on the walk side, an edging of the rustic irons elsewhere described. As soon as they get beyond the very primitive idea, that because one border is of a certain pattern the others ought to follow it, this will be found a really good plan, and it is worth attention with us; by its means we may enjoy great variety in a border without any of the raggedness of the old mixed border system. Around most of the rose trees they place a ring of the Gladiolus—a good plan where the plant grows well. Any person with a knowledge of bedding plants may vary this plan *ad infinitum,* and produce a most happy result with it wherever borders have to be dealt with. In the Tuileries we find much the same kind of garden, with a fine bloom of roses in summer, and more hollyhocks and Althæas in autumn. So also in the Palais Royal we meet with the same style going around the squares, but it is only the Louvre strips that have the wide margin of dark green Irish ivy—a refreshing and neat finish. It is not far to step into the garden of the Palais Royal, but there is little to be seen there which we shall not meet elsewhere. There are great crowds sitting out of doors in the evenings; a smaller crowd at noon waiting to set their watches by the little bomb which is fired by sunlight every day at that hour, but in the plant way I only remember it for a few specimens of the round-headed Acacia inermis in tubs, furnishing as good an effect as many things more difficult to grow and obtain.

THE TUILERIES.

Let us next go to the west end of the palaces to see the gardens of the Tuileries, which stretch from the western face of the palace to the Place de la Concorde, bounded on one side by that fine straight street the Rue de Rivoli, on the other by the river. Being nearly in the centre of Paris these gardens are the most frequented of all perhaps. The garden is very large, and laid out in the plain geometrical style by Le Notre, wide straight walks, borders round grass plots dotted with little lilac bushes, and flowers below them. About one-fourth of it near the palace is cut off for the Emperor's private use, but this part is merely divided from the public one by a sunk fence and low railing, so that the view of the private garden is enjoyed by all. In it they simply plant good evergreens and plenty of deciduous flowering shrubs, while the grass plots are belted by borders, and one runs right along under the palace windows with the usual round bushes of lilac, but these borders are kept pretty gay all the year round. The private garden of the Emperor is quite open to the public when he is not living at the Tuileries. It is well worth visiting should an opportunity occur, if only to see the way the ivy edgings are used. There are no beds, only borders—these touching the gravel walk, and being edged with box. Then on the bright gravel itself, or apparently so, they lay down a beautiful dark green band of the Irish ivy, of course allowing in the laying down of the walk for the space thus occupied. The effect of the rich green band adds much to the beauty of the borders. The flowers are kept a good deal subdued, and some trouble is taken to develope the shrubs and stronger vegetation distinctly and well. The effect is very good from the windows and the interior. Here again we see the ivy edging, twenty inches wide, well used ; and Cannas afford a very charming effect in mixed borders. A very wide walk crosses the garden just outside the private division ; at about its centre are a large basin and fountain, from which another wide walk goes straight towards the Place de la Concorde, and by looking in that direction you see the whole length of the magnificent Avenue des Champs Elysées, terminated on the crest of the hill by the Arc de Triomphe. This walk cuts the garden into two portions chiefly planted by chestnuts and other tall trees, which have not been sufficiently thinned, but are allowed to run up very tall, and thus afford a high arched shade in summer, the ground being gravelled underneath so that it is comfortable to walk or play upon. There is a slight narrow terrace on both sides, an orangery, the contents of which are placed out in summer, an alley arched over with lime trees by the side of the Rue de Rivoli, and at the western end there are terraces which afford a capital view of the bright and busy scene around and the magnificent avenue towards the west. The sculpture is good, and there is a great deal of it, both copies of celebrated works and original ones, but as for fresh horticultural interest there is little or none to be seen ; and a passing glance is all the visitor need bestow on the public part of the garden of the Tuileries, though it is only fair to add that its general effect is very good, and that it in all respects answers its purpose as a play and promenading ground, and a " lung " to the city.

But let us wait a moment to look at these people feeding the birds so much to their own amusement, and also that of the lookers-on. It is a pretty sight, and seems to afford much pleasure to many people, and doubtless much more to the successful feeders. It is quite a little scene in the gardens every day, and on fine days it attracts numbers of people, though it is an every-day occurrence there. The Jardin des Tuileries is inhabited by a great number of the common " ring dove," or " quest"—those wild pigeons which in Britain and elsewhere, when in a wild state, flash away from man like an arrow from the bow. In these and other gardens in Paris they seem perfectly at home, and perch at ease in the trees over the heads of the multitudes of children who play, and of people who walk on fine days. Their intimacy does not extend further except with their friends who come to feed them now and then. Here is an instance. A man, evidently a respectable mechanic, comes to a certain spot, near the private garden of the Emperor. Presently some of the pigeons fly to their friend. He is an old acquaintance, and a bird alighting on his arm gets a morsel of bread to begin with; others follow. He has previously put a few crumbs of bread into his mouth, of which the birds are well aware, and, arching their exquisitely graceful necks, put their bills between his lips and take out a bit "turn about." Perhaps one alights on his head, and he may accommodate two or three on his right arm. There are others perched on the railings near at hand, and they come in for their turn by-and-by. A dense ring of people stand a few yards off, looking on, especially if it be a fine day, but they must not frighten the birds, and this persistent feeder looks daggers at a small boy who allows an audible yell of delight to escape. Presently the sparrows gather round the feeder's feet, and pick up any crumbs that may fall while he is transferring the bread from his pocket to his mouth. The sparrows, sagacious creatures, do not as a rule light upon the arm, and never even think of putting their heads in the mouth of the man, but flutter gently so as to poise themselves in one spot about fifteen inches or so from the hand of the feeder. He throws up bits among them, and they invariably catch them with slight deviation from their fluttering position, or at most with a little curl. It is very pretty to see them thus fed, and to see the exquisitely graceful heads and necks of the wood pigeons as they move them to extract the crumb is charming. In one instance we saw a sparrow or two alight on a man's hand, and pluckily root out crumbs that he held rather firmly between his finger and thumb. He was an ancient and persevering personage, evidently of the Jewish persuasion; and however much I may regret to admit it, as a faithful chronicler I must state that not one sparrow approached within ten inches of the hand of a Gentile. Similar instances of this interesting bird-feeding would be a pleasant pastime in other places than the Tuileries gardens.

The Champs Elysées, &c.

The Place de la Concorde is not a garden but a noble open square—a worthy centre to the magnificent roads and streets that are near and radiate from it, and with a terrible history. Fountains and statues embellish it; there is no sign of its fearful associations to be seen; it is a place to make one ashamed of Trafalgar-square, which has been ridiculously called the finest site in Europe. By looking to the east the Palace of the Tuileries may be seen through the division made in the wood of chestnuts by the central walk, and to the west the Avenue des Champs Elysées. If the reader who has not visited Paris will suppose the lower part of Regent-street to be flanked with a large and wide pleasure ground, with a grand tree-bordered avenue passing through its centre straight away to the highest point of the broad walk in the Regent's Park, and there crested by an immense triumphal arch—the largest in the world, 161 feet high and 145 wide—he will be able to form some idea of what the scene here is. It was only in 1860 that the garden part was laid out as such, and yet it looks an ancient affair, has many respectable specimens of conifers, Magnolias, &c., numerous large and well made banks and beds of Rhododendrons, Azaleas, hollies, and the best shrubs generally, with abundant room for planting summer flowers, chiefly however as margins to the clumps of fine shrubs. It is a large space, large enough for the allied armies to encamp here in 1813, and perhaps the most remarkably diversified scene of the kind in the world. In the centre is the avenue itself, usually covered in fine weather with carriages; then wide asphalte footways and gravel walks, with about five rows of trees on each side, and then the ground begins to assume the gardenesque aspect, and to become in short a first-rate pleasure garden. Here and there among the trees may be seen neat and showy-looking buildings surrounded by a close planting of rhododendrons or other shrubs, and with a rather large gravelled space surrounding each; these are cafés, where singing, music, &c., are to be heard in the summer—the summer residences of the music halls in fact. Other "places of amusement," from Punch and Judy shows to revolving circuses, abound. On Sundays and fine days the wide tree-shaded walks are crowded with pedestrians; all the little games are in full swing, and though it may seem a queer jumble to some readers, the whole thing is about as orderly as a crowd at a flower-show. A little above this is the Rond Pont, a circular open space, where a lot of avenues radiate, and in which there are large beds for flowers, fountains, &c., disfigured, however, by the undulations which some poor little bits of grass are made to assume. Object-less and unnatural diversification of the ground and the lumping together of too many things in one mass, are the weak points in the gardening of Paris. Surely a better use could be made of the beautiful variegated Acer negundo than putting many score plants of it in one great pudding-like bed! They do the same thing to some extent with the subtropical plants, but it is a fault which must be cured some day in the natural order of things. The best way to go to the Bois de Boulogne from the Arc de Triomphe is along the grand Avenue de l'Impératrice, which is bordered by grass, trees, conifers, fine villas, gardens, &c., and is in itself worth seeing. It leads straight to the Bois, the most fashionable of Parisian parks. This should be seen by everybody; and as La Muette at Passy, the great propagating establishment spoken of elsewhere, lies on its east side, there is no lack of horticultural interest in the vicinity.

The great arch is surrounded by an immense circular space, from which straight boulevards and avenues radiate in all directions. The guide-books advise the visitor to Paris to see the lamps lit at night in the Champs Elysées, but if he should want to see the finest effect of that kind, he must go to this arch of a dark night, and standing in the centre look at their effect in the great avenues, which fall from where he stands, and afterwards walk round the arch to see them better still. The whole scene around here is magnificent, and highly suggestive of how a great city should be laid out. If Paris had nothing worth seeing but what may be seen from hence, it would well repay a visit to all persons interested in the improvements of towns and cities. One of the avenues that radiate from this great circle is the Avenue de la Reine Hortense, and at its end may be seen large and handsome gilt gates; they are at the entrance to the

PARC MONCEAU.

This is a place which should be seen by every horticultural visitor to Paris. It is not large, but exceedingly well stored, and usually displays a vast wealth of subtropical plants during summer. In spring it is radiant with the sweet bloom of early-flowering shrubs and trees, and every bed and bank covered with pansies, Alyssum, Aubrietia, and all the best known of the spring flowers, while thrushes and blackbirds are whistling away as if miles in the country, though it is only a few minutes' walk from the Rue du Faubourg St. Honoré. It contains abundant plantations of good shrubs, large trees, with shady walks through them, and masses of one kind of shrub here and there. More than 150 plants of the fine Acer negundo variegata may be counted in one clump, and around nearly all the clumps are detached circular little beds, filled with sandy peat, just big enough for one plant. A lot of specimens of Thuja aurea outside one clump of trees and shrubs is a handy illustration of the way they dot about the subtropical plants in summer ; and, by the way, if we made good use of such plants we should have little want for costly " subtropicals." In one large bed Aralia papyrifera was left out all the winter, the bed snugly thatched over ; but though the plant may be kept in that way, it does not seem a desirable course to pursue. This park was laid out so long ago as 1778 for Philip Egalité as an " English garden," and passed through various changes, till it at last fell into the hands of the Municipality of Paris, a very astute corporation, who have converted it into a charming garden, and are not likely to part with-it in a hurry. It contains a small and not pretty lake, half encircled with round fluted columns.

Dr. Moore, of Glasnevin, once told me that he considered this in its full dress the most successful example of flower-gardening he had ever seen, and therefore it may be well if we look at it in that state.

The system of planting adopted here as well as in the other gardens of the city is often striking, often beautiful, and not unfrequently bad. It is striking when you see a number of that fine silvery tree,

Acer negundo variegata, arranged in one great oval mass, gay and bright; it is beautiful when you see some spots with single specimens and chaste beds, every one differing from its neighbour; and it is bad indeed when you meet with about a thousand plants of one variety stretched around a collection of shrubs, or flopped down in one wide mass in some of those beautiful little green pieces that spread out among the trees on the islands in the Bois de Boulogne. In the Champs Elysées there is a deal of this gardening, and much of it consists of masses of monstrous ugliness, in consequence of what we may call overlumping, to produce sensational effect; but it is less observable and more artistically managed in the Parc Monceau. An odd single specimen of Musa Ensete looks very fine here and there, and occasionally you may see a really tasteful thing—for instance, a mass of brightly and freely flowering Portulacas surfacing beds with tall foliaged plants, &c. The Hibiscuses begin in August to compensate for months of ugly stickiness by showing an odd blazing bloom, beautiful here and there, while Begonias and not a few other things look miserable indeed. "This subtropical system will never do for England!" say some practical men. The truth is, that it requires to be done very carefully in Paris, and there is a great mistake made by putting out a host of tender plants merely because—well, I can't say why, unless it is to contrast healthy beauty with ragged ugliness. Their weakness is, that like most people in this world, they do not know where to stop. You have in the Parc Monceau a group of Musa Ensete worth making a journey to see, and groups of Wigandia, Canna, and such Solanums as Warcewiczii, that are worthy of association with it; but you also see there beds of Begonias without a good leaf, let alone a particle of beauty: nasty scraggy stove plants, with long crooked legs, and a few tattered leaves at the top, and poor standard plants of the sweet Verbena at the same time. If it were an experimental ground one would not mind, of course; but this, in gardens where its omission would leave almost nothing to be desired, is almost too bad. In some respects this park is really unequalled, and therefore one regrets the more to see these blemishes. However, it is only fair to say that last season was exceptionally bad, and that many of the plants and much of the skill of the chief gardeners of the town had to be devoted to the grounds of the Exposition.   Musa rosacea and M. sinensis flap their poor leaves about in the wind without being "kilt entirely," but among all the Musas, M. Ensete is the best for the open air or the cool conservatory.   I have seen it withstand hailstorms without suffering in the least.

Such plants as Crambe cordifolia and the Pampas-grass look fine isolated on the grass near the margin of a clump of trees.   The Pampas is worth growing for this purpose alone, even if it never flowered.   Erythrinas flower well, but should not be used in large masses.   Beds and borders of Hydrangeas are very fine indeed, and worth more attention than they get in England. Some plants bearing blue flowers amongst the normally coloured ones, of course add much to the effect.   An immense mass of Canna nigricans, with edge of variegated Ageratum, was very imposing.   Clematis montana trained up the trees has a charming effect.   What a nice thing it must be in flower in this position! though it would be better allowed to run over some old stumps or low common trees over which it might train itself.   Beds of ferns in shady places are nice, and lines of white Fuchsias are pretty. Ferdinanda is good.   A little way off, single plants of Erythrina look very striking.   Some tall slender Solanums, &c., are quite a disfigurement.   Colocasia odorata is very fine, and free and noble, especially when the plants are old and furnished with tall stems. Isolated plants of Bambusas are beautiful and striking, as are also those of Acanthus.   Delphinium pegged down amongst Phloxes is very good indeed; Arundo Donax versicolor is used, but it is very poor, in consequence of not having been left in the ground over the winter.   Hibiscuses are too naked to be anything but ugly, and, no matter what result they may produce in the end, no such hideousness as they present for a long time should be tolerated in any tasteful garden.

Beds of the better kinds of Solanums are very fine; but I need not enumerate the plants that furnish the best effect, inasmuch as these are fully treated of in the portion of this book devoted to subtropical gardening. The source of the superior beauty of this garden is the *variety* which it presents in all its parts, even the great masses of Canna and the variegated Negundo are not presented in duplicate.

The Bois de Boulogne.

This park illustrates how we improve by friction, so to speak. Till 1852 the Bois was a forest, a dwarf tangled one too, for when the English and Prussians encamped in it in 1815, they cut down everything for fuel. But Napoleon III., in his admiration for English parks, determined to add their charms to Paris, or rather to improve upon them, and the Bois is one result. In concert with the municipality, the Emperor dug out the lakes, made the water-falls, &c. As a combination of wild wood and noble pleasure garden, it is magnificent. The deer are placed in a closed-off space. The Bois is splendid too as regards size—much larger than the Phœnix Park at Dublin; in fact, nearly 2500 acres. Though with large expectations in other directions, the reader will hardly be prepared for the statement that the French beat us in parks. When first entered it may not be much liked, the numerous Scotch pines around one part of the water giving it a sort of pine-barren look, but a few miles walk through it soon dispels that idea. It has more than the beauty and finish of any London park in some spots, but on the other hand, vast spreads of it are covered with a thick, small, and somewhat scrub-like wood, in which wild flowers grow abun-dantly, unlike the prim London parks. There were plenty of wild cowslips dotted over even the best kept parts of it last spring, and groups of those splendid deciduous Magnolias could be seen for long distances through the then leafless oaks. To see what the Bois de Boulogne really is, the visitor should keep to the left when he enters from Passy or the Arc de Triomphe, and go right to the end of the two pieces of ornamental water. Then, standing with his back to the water, he will notice an elevated spot with a cedar growing on it, and by going to that spot he will get the finest view he has probably ever seen in a park—the water in one direction looking like an interminable inlet, beautifully fringed with green and trees, and in the other several charming views are opened up, showing the hilly suburban country towards Boulogne, St. Cloud, and that neighbourhood. Then, by turning to the right and coming again to Paris by the other side of the water, he will have a good idea of what a noble promenade, drive, and garden this is. It is in all respects worthy of its grand approaches, of the width and boldness of which people connected with English parks and gardens, and who have not seen those of Paris, can have no conception. There is some bold rockwork attempted and well done about the artificial water; and very creditable pains are taken to make the vegetation along it diversified in character, so that at one place you meet conifers, at another rock shrubs, in another magnolias, and so on; and not the eternal repetition of common things which one too often sees at home. If you have time, go across the Bois as far as Longchamps, not for the sake of seeing the race-course, which attracts half Paris to this part of the wood on fine Sundays, but a large and beautiful cascade which faces it. Above the "spring" or shoot of the large cascade is an arch of rustic rocks, over which fall ivy and rock shrubs, the whole being backed with a healthy rising plantation. The fault of the most gardenesque part of the Bois de Boulogne is that the banks which fall to the water are in some parts a little too suggestive of a railway embankment, and display but little of that indefiniteness of gradation and outline which we find in the true examples of the real "English style" of laying out grounds. But you do not notice this from the position above described, from whence indeed the scene is charming. The fault just hinted at is common to almost every example of this style to be seen about Paris; and in most of their walks, mounds, and the turnings of their streams, you can detect a family likeness and a style of curvature which is certainly never exhibited by nature, so far as we are acquainted with her in these latitudes. But it is only justice to say that, taking the parks on the whole, they are far before our London ones in point of design. We have nothing that approaches the Bois de Boulogne—nothing that surpasses some parts of the Bois de Vincennes.

Of the Pré Catelan, and certain minor features of the Bois de Boulogne, space forbids me to speak ; and I must omit all notice of the Bois de Vincennes, which is well worthy of a visit, and exhibits some capital arrangement of artificial water.

THE GARDEN OF THE ACCLIMATIZATION SOCIETY IN THE BOIS DE BOULOGNE.

This is a pretty garden and a most interesting place. In it you may study many things, from the culture of the oyster to the numerous breeds of domestic fowls, from the various silkworms to the different plants used for bee feeding. I took a deep interest in the hybrid ass—a neat cross between the domestic and wild varieties, albeit he proved useless for the carriage, and kicked it and the harness into "smithereens" when yoked, in consequence of the virus, or what an Irishman would call the "divilmint" of the exotic parent predominating ; I was not insensible to the claims of a Russian dog with a coat like a superannuated door-mat ; I laughed at a duck which had a velvet-looking head remarkably like a hunting-cap, and nearly as big, but with a body no larger than a debilitated blackbird ; and was amazed to see a *chien* with no hair except on the top of his head. But we must let all these things pass, and confine ourselves exclusively to vegetable life—always of great importance, since man first regaled himself upon fruits and green-meat. Doubtless one of the first things the sagacious creature pitched upon was the grape—at least the best varieties of grapes and the best varieties of men are supposed to have originated in much the same place. To-day it is more important than ever, and the gardeners here were employed in planting the last of a magnificent collection of 2000 varieties of vines ! What a grand collection ! It is the famous one formed in the gardens of the Luxembourg, and fortunately saved from destruction by M. Drouyn de Lhuys, acting upon the urgent request of a friend of horticulture.

They were actually about to be thrown away when the recent mutilation of the Luxembourg garden took place. So by authority they were ordered to the gardens in the Bois de Boulogne, where, let us hope, they will be well looked after, as it would be a great pity if a collection embracing, as far as could be gathered, nearly all the varieties cultivated in the world, should be lost to horticulture and to science. By the way, I observed where they were digging deep for sand in this garden that it is placed on a bed of capital garden gravel almost as thick as the bed of dense clay under the Zoological Gardens in the Regent's Park, and such a basis is of course excellent for a thing of the sort. I saw a man carrying manure on his back to the vines, and sat down and contemplated him going through the interesting task, the basket being placed on a slightly elevated board supported by three sticks, from which he could readily hook on to it when filled. I looked at him with respect, and some sympathy, just as we should at a living specimen of the Dodo or any other animal supposed to be extinct. It occurred to me at the time that the acclimatization of a handy useful species of wheelbarrow would not be unworthy of the Society. However, it is only fair to add that this kind of basket or *panier* would prove useful in town gardening, as soil has often to be carried through the house and also for carrying vegetables, &c.

The Lycopodium is used with charming effect to form a turf in the conservatory, and nothing can look better than the New Zealand flax, and several palms and tree ferns, planted near the margin of a winding bit of water in that structure. Musa Ensete too looked nobly in the same position, planted out in a coolish house. Although the glasshouses in the garden afford but little interest, rockwork and the planting out of fine foliage plants tend to make the conservatory very chaste and refreshing. Those who visit it during the winter, cannot fail to be much struck with the effect produced by beds cut in the rich green of Lycopodium denticulatum, and filled with Primulas, Cinerarias, and spring flowers generally. The whole floor of the house, walks excepted, was effectually covered by the Lycopodium.

THE LUXEMBOURG GARDEN.

This beautiful old garden, the favourite resort for many years of the Parisians of the left bank, has lately been sadly pulled about, much to the indignation of the Parisian public and journalists; but it is still a pretty garden. Geometrical gardens are rarely capable of affording any prolonged interest or refreshing beauty; very rarely so much so as that of the Luxembourg. Before the recent alterations there was a good botanic garden—an irregular sort of English garden, which the French call the "never to be forgotten nursery" —and much miscellaneous interest now past away. At present matters are much more concentrated, and we shall find far less to speak of than of old, but yet enough to make the place worth a short notice.

The graceful way in which the French use the common ivy is pretty well exemplified in these gardens near the fountain built by Catherine de Medicis. Stretching from this fountain there is a long water-basin, a walk on each side of that bordered with plane trees, which meeting overhead make a long leafy arch, so that the effect of the fountain group at the end, representing Polyphemus discovering Acis and Galatea, is very fine. Its effect is of course heightened by the leafy canopy of planes, but very much more so by the way in which the ivy and Virginian creeper are made to form graceful wreaths from tree to tree. Between each tree the Irish ivy is planted, and then trained up in rich graceful wreaths, so as to join the tree at about eight feet from the ground. At about a foot or so above the ivy another and almost straight wreath of Virginian creeper is placed, and the effect of these two wreaths from tree to tree is quite charming. The wreaths seem to fall from the pillar-like stems of the plane-trees rather than to grow from the space beneath them, the bottom of the lower wreath resting on the earth. An adoption of this plan would add verdure and grace to many a formal grove, bare and naked-looking about the base Another thing right well done here during the past year was the arrangement of vases. Instead of lumping them with flat round-

headed subjects, as many do with us, they placed in the centre of
each a medium-sized plant of the New Zealand flax, with its long
and boldly graceful leaves, and then set geraniums, &c., around,
finishing off with the ivy-leaved geranium, and the Tropæolum for
drooping over the margin.

In these gardens the Oleander is grown into large bushes like the
Orange-trees, and put out with them during the summer months
They become perfect beds of flowers. I have seen plants or rather
trees of those oleanders in flower here, quite ten feet through, and
with the flowers as thick upon them as on a bed of geraniums.
They are simply treated like the orange trees, the culture of which
is fully described elsewhere in this book. We do not know if the
plan would succeed in England, but it is worth a trial. The
Oleander is not often flowered well with us, though quite worth the
trouble of cultivation.

On the 5th day of July they were busily employed moving large
chestnut and plane trees in full fresh leaf in these gardens. They
take them up with immense balls and powerful machinery, and very
successfully, but it is not a system that should be pursued more than
is barely necessary in private gardens. It may be very desirable for
La Ville de Paris to move ordinary trees of goodly size to complete
and rearrange its straight avenues here and there, but it is a thing
that should be attempted as little as possible. Such trees are not
worth the expense in any other case. The roses used to be very
good about the Luxembourg, and they are so still, though the fine
old rosary is done away with. The grass banks that surround the
geometrical garden—such slopes as may be seen in most places of
the kind—are not left naked, but planted with two rows of nice
dwarf rose bushes, and the effect of these is very charming. There
seems no particular reason why like spots should be left naked with
us. Let us pass by the fine collection of orchids, camellias, &c., and
some other interesting features, and come at once to the style of
decoration in summer, which is distinct, good, and well worthy our
attention. Continuous borders, not beds, run round the squares of
grass, &c., and from the dawn of spring to the end of autumn, these

are never without occupants, never ragged, never flowerless. It is not the old mixed border system, observe—far from it; nor is it the bedding one. It is a system of bedding and herbaceous plants mixed, but all changed every year. They steal out a spring flower this week, and put in a fine herbaceous or bedding plant, or strong growing florists' flower in its stead, and with the very best success. Stocks of good bedding and herbaceous plants are always kept on hand to carry out this, and the placing of the herbaceous plants into fresh ground every year causes them to flower as freely as the bedders. But those borders also contain permanent things— lilac bushes, roses, &c., which give a line of verdure throughout the centre of the border, and prevent it from being quite overdone with flowers. Among those woody plants there were others very beautiful and very sweet for many weeks through the better part of the season, and those were low standard bushes of the common honeysuckle! English gardeners would perhaps scarcely ever think of that for such a position; but alternating between a rose and a lilac, or other bush, and throwing down a head of free-growing and flowering shoots, very few subjects look more pleasing in the flower garden. Few arrangements can be more satisfactory than the mixture of Phloxes, Gladioli, Œnothera speciosa, Fuchsias, Pelargoniums, large yellow Achillea, &c., in these beds at present and for months past. They also have the subtropical system at the Luxembourg, and rather more tastefully than elsewhere. Thus in one part may be seen a graceful mixture of a variety of fine-leaved plants with an edging of Fuchsias, instead of the ponderous mass of 500 plants of one variety of Canna, which you sometimes meet with in other places about Paris. M. Rivière is fond of having mixed beds of ferns in the open air, and isolated specimens of tree ferns, Woodwardias elevated on moss-covered stands, &c., and their effect is very good.

Numerous amateurs and others go to the Luxembourg to hear M. Rivière deliver his free lectures, which are thoroughly practical, and illustrated by the aid of living specimens and all the necessary material. The lecturer goes through the theory and practice of the matter before an attentive class, consisting of several hundred persons, and elucidates the subject in a way which cannot fail to highly benefit the numerous amateurs who attend the classes. It is a very interesting sight to see such a number of people here at nine o'clock in the morning, taking a deep interest in the matter, and speaks much for the excellence of the professor. There are many lectures delivered in England on like subjects, but none so directly useful to the horticulturist as these.

Parc des Buttes Chaumont.

This is the boldest attempt at what is called the picturesque style that has been attempted either in Paris or London. For my own part I have an opinion—it may be a weakness—that attempting expensive and extraordinary works in places of this sort is not wise, at least till all the densely populous places are provided with healthy well-planted parks. Thus I think that in London it is a mistake to devote great expense to a few parks, and leave so many square miles of population without a green spot. But in this instance an unusual attempt was to some extent invited by the existence of a great quarry in one part of the ground. The whole park may be described as a sort of diversified Primrose Hill. Imagine that, with two or three "peaks and valleys," and with an immense pile of rock seen here and there, and you have a good idea of this park. At its hollow or lower end there was a quarry, and this has been taken advantage of to produce a grand feature. They have cut all round three sides of this quarry, smoothed it down, leaving intact the great side of stone, and adding to it here and there masses of artificial rock. This forms a very wide and imposing mass of rock, 164 feet high, or thereabouts, in its highest parts, and from these you may gradually descend to its base by a winding rough stair, exceedingly well constructed, and winding in and out of the huge rocky face. At the base of the cliff, and widely spreading round it, there is a lake. Then this ponderous cliff—and it is best described by that name—has several wings, so to speak, and in one bay is made a most effective stalactite cave—enormous is the right word, for from its floor to the ceiling is about sixty feet high, while it is wide and imposing in proportion. In its back part above this high roof the light is let in through a wide opening, showing a gorge reminding one of some of those in the very tops of the Cumberland mountains, and down this gorge trickles the water into the cave, ivy and suitable shrubs being planted along its course above the roof of the cave; and the effect is remarkably striking, though it is hardly the kind of thing to be recommended for a public park. By all means let us leave the luxuries of gardening out of the question, till we have provided the necessaries for the population of great towns, and these are green lawns, trees, and wide open streets and ways, with their necessary consequence, pure air. In one of the buttes, or great mounds here, they have planted 500 or 600 Deodars—made it a hill of deodars in fact. This is a mistake, for though Paris is not as foggy as Spitalfields, nor its air as destructive to trees as that of the West-end of London, it is a great city, as may be seen from this park, and with many a vomiting chimney too, so that the better plan would be to pay double attention to deciduous trees, using only such evergreens as are certain to grow. In one wide nook, perfectly sheltered on the three coldest sides, M. André has planted a collection of subjects mostly tender in the neighbourhood of Paris. From this park, the surroundings of which are by no means nice-looking, you can look over nearly all Paris. The approach to it from the central parts is shabby for a Paris approach, and on your way you may catch some idea of what Paris was before the splendid improvements of the past ten years—a very dirty city; but this approach, like most other things in Paris, is simply tolerated till more important things are finished. Of the quick way in which they turn them out of hand there, the reader can scarcely have a notion. I have seen acres of land removed bodily without any fuss being made about it; miles of trees planted in the course of a single week; old suburbs blown up by hundreds of mines a day, and levelled into commanding terraces fit for princely mansions.   One June day, bright, dry, and very warm, they were planting trees in this park, and large ones too—trees that required great machines to lift them—while they were marking the ground for fresh plantings.   Do you plant after this date? I asked.   Yes, every day in the year !   There is an imitation of a small Roman temple erecting on the summit of the great cliff, while the house of the guard of the park, the entrance lodge, the bureau, and the refreshment rooms, are in neat cottage style, and so placed as to add much to the appearance of the park, which in the more finished spots looks like a well-cared-for bit of green England.   They are making many beds of peat for rhododendrons, and their peat is of an admirable silvery kind, much admired by all the English who have seen it with me.   Their late planting, however, is not to be recommended, and, as its consequence, many of the young trees look as the accomplished Kay of the casual ward described one of the performers at the " Victoria "—" wery dicky." This park has some very unusual features indeed, and should be seen by every visitor to Paris, being more remarkable than others better known.

The Jardin des Plantes.

We have nothing in the British Isles like the Jardin des Plantes. It is half zoological, half botanical, and nearly surrounded by museums, containing vast zoological, botanical, and mineralogical collections.   The portion entirely devoted to botany is laid out in the straight, regular style, while the part in which are the numerous buildings for the wild animals, has winding walks, and some trifling diversity here and there.   The place is really an important school of science, and as such it is great and useful.   In addition to able lecturers on botany, culture, and allied matters, there are, I believe, a dozen lecturers on various other scientific subjects, some of these gentlemen being among the ablest and most famous naturalists in Europe.   Here Buffon, Cuvier, Jussieu, and many other leading men have worked; and here at the present day, even in minor departments, are many men of well-known ability.

Although the Jardin des Plantes is quite inferior in point of
beauty to any of our large British botanic gardens, it contains
some features which might be introduced to them with the greatest
advantage. Its chief merits are that its plants are better named
than in any British garden; it possesses several arrangements which
enable the student to see conveniently, and most correctly, all obtainable useful plants infinitely better than in any British botanic
garden; and it displays very fully the vegetation of temperate and
northern climes, and, consequently, that in which we are the most
interested, and which is the most important for us. Its chief faults
are that it has a bad position in an out-of-the-way part of the town;
the greater part of its surface is covered with plants scientifically
disposed; the houses are poor and badly arranged compared to
those in our own good botanic gardens; and there is no green turf
to be seen in its open and important parts. It has, in addition, a
very bad "climate" for pines and evergreens, and there is a ridiculous
kind of maze on the top of an otherwise not objectionable mound.
Half way up this elevation stands a tolerably good Cedar of Lebanon,
the first ever planted in France. It was planted by Jussieu, to whom
it was given by the English botanist Collinson. Beyond this there
is not much tree-beauty or tree-interest in the Jardin des Plantes.
The houses are in some cases very well cared for, but are inferior
to those in English gardens. There are fine collections of Palms
and other subjects of much importance for a botanic garden, and
the house collections are on the whole good, but the plants in a
great many cases are very diminutive and poorly developed, therefore we will pass them by.

There is one admirable feature which must not be forgotten, and
that is the magnificent collection of Pear trees. M. Cappe has had
charge of this section for about thirty-five years, and is now a very
old man, but still he attends to his trees, and has them in fine condition, though contending with much difficulty, because the space
upon which the trees stand is really not enough for one half the
number, and thus he is obliged to keep lines of little trees between
and under big ones, and so on. There are few things in the

horticultural way about Paris better worth notice than this collection of pears.

Remarking that they have a graceful way of commemorating great naturalists by naming the streets in the immediate neighbour-hood of the garden after them, I will pass on to the more important feature of the garden; that is, its very extensive and well-named collection of hardy plants.

The only species of Pelargonium that ventures into Europe (P. Endlicherianum) is grown here, and it is quite hardy. Space pre-vents me from naming the many interesting hardy plants seen here and elsewhere about Paris, but I do not intend to forget them, and have already introduced some of the most important.

The first of the principal arrangements of these plants is a curious and distinct one. It is simply two large and wide spaces planted with masses of first and second-rate species; and looks pretty well, though far from being arranged in a way to develope fully the beauty of its contents. Edgings composed of the several varieties of Iris pumila, looked very well in early spring, and many plants are used for edging which we are not accustomed to see so employed in England. Thus the good double variety of Lychnis Viscaria has been very pretty as an edging, and so has the neat, bright, and pure white Silene alpestris, an alpine plant not half so popular as it ought to be, though I observe that some seedsmen, while not offering it, vend a pretty free proportion of the weeds that belong to the genus.

Then there is a large space devoted to plants used for the deco-ration of the parterre, all or chiefly tender plants or annuals. This is not so successful or useful as some of the other arrangements. Let us pass on to a large division devoted to the culture of plants used as food, and in commerce. It is at once successful, useful, and complete. The chief varieties of all garden crops, from Radishes to Kidney Beans, are to be seen; the various species of Rhubarb, all important varieties of Lettuce—in a word, everything that the learner could desire to see in this way. It is not merely the plan of the thing that is sensible and good, but its carrying out.

Here the annuals are regularly raised and put out; the ground is kept perfectly free from filth, and it is, in fact, the best place I have ever seen to become acquainted with useful plants. Such arrangements well done, and hidden off by judicious planting from the general verdant and chief area of any of our great public gardens, would be of the greatest service. The ground is thrown into beds about six feet wide, and each kind is allotted six feet run of the bed. The sweet potato is grown here, as indeed are all interesting plants that may be grown in the open air. Such readers as care about this root, which by the way is of agreeable flavour when well cooked, may grow it most readily and effectively by placing it in a frame or pit after the spring crop has been taken out; or, indeed, on a ridge like the ridge cucumber; but the pit or frame is the safest way generally—the lights being taken off. As pits and frames are frequently empty from about the 1st of June till autumn, room might be readily spared for it without loss, and a useful vegetable added to our stock, which is yet in want of variety, fine as it is. The roots may be bought in Covent Garden. The red variety is the best. The way to treat them is to pot them about the end of April; start them in a gentle heat, and have them fresh and stubby for planting out in the pit or frame about the 1st of June. They would be the better for the lights for a few days. In this way they will be found to do better than when grown in a stove, and probably prove a more grateful vegetable than the Chinese yam in its best state. Below this arrangement, and near the river end of the garden, is another very interesting division. It is chiefly devoted to medicinal and useful plants of all kinds, arranged in a distinct way. First we have the Sorghums, Millets, Wheats, and Cereals generally —all plants cultivated for their grains or seeds. Then come plants cultivated for their stems, from Polymnia edulis to Ullucus tuberosus. Next we have the chief species and varieties of Onion, such plants as Urtica utilis, the Dalmatian Pyrethrum rigidum, and in a word almost everything likely to interest in this way, from Lactuca perennis to the esculent Hibiscus. Here again the plants are well named and kept clear and distinct, each having full room to develope,

the general space devoted to the subject being sufficiently large.
All these divisions we have just passed through cover an oblong
expanse of ground, the effect of which is of course anything but
beautiful from an ornamental point of view, but yet, in consequence
of the ground being well kept, each subject grown well and
vigorously, and all the squares bordered with roses and summer
flowering plants, the effect is better than might be expected.   This
great oblong space is bordered on each side by double rows of lime
trees planted by Buffon.   Between these are wide walks, agreeably
shady on hot days.

The second great oblong space to the north is entirely devoted to
the school of botany, and this is simply a large portion of ground
planted on the natural system, remarkable for the correctness of its
nomenclature, and the richness of its collection.   Here again
everything is well taken care of and kept distinct ; the aquatics are
furnished with cemented troughs, in which they do quite luxuriantly ; and the whole is most satisfactory, with one exception—
that they place out the greenhouse and stove plants in summer to
complete the natural orders.   These poor plants are stored pell-mell
in winter in a great orangery, from which they are taken out in
early summer literally more dead than alive.   They make a few
leaves during the summer, and are again put into their den to sicken
or die.   The medicinal and other plants for special uses are indicated by variously coloured labels.

For the information of curators of botanic gardens I may state
that Cuscuta major is luxuriantly grown here upon the nettle,
C. Epithymum upon Calliopsis tinctoria, C. Engelmanii upon a
Solidago, and Orobanche grows upon Hemp.   I have grown
O. minor upon perennial Clovers, and O. Hederæ may be readily
grown upon the Ivy at the bottom of a wall (I once saw it growing
freely on the top of a wall near Lucan, in Ireland), so that there
ought not to be the difficulty in showing these curious plants to the
public which our botanic gardeners find.   Orobanche ramosa is also
grown here upon Calliopsis tinctoria.   The safest way with the
Orobanches is to scrape away the soil till you come near the root of

the plant on which you intend it to be parasitical, and then sow the seed. Juncus glaucus may be seen here and in numerous French gardens planted near the water : it is very extensively used in nearly all French gardens for tying plants. They plant a ring of it round the water cistern, or any convenient spot, and cut a handful whenever it is required. It seems a useful practice. Fruit growers like the rush for tying down the young shoots, while it is also useful for tying up annual and herbaceous plants.

Should any visitor to the Jardin des Plantes wonder at the poor external aspect of its houses and some other features as compared with those at Kew, he would do well to bear in mind that money has a good deal to do with such things ; and that the grant for museums, lecturers (the lectures are free), the expensive collection of animals, and everything else in the Jardin des Plantes, is miserably small. On the other hand, the gardens and plants of La Ville de Paris are luxuriously provided with the "needful ;" the municipality of Paris often spending prodigious sums for the purchase of plants, and even for the plant decoration of a single ball. One ball at the Hôtel de Ville during the festivities of the past year cost considerably over £30,000, while the poor Jardin des Plantes gets from the State—I am not quite sure how much, but probably not one-third of that sum, to exist upon for a whole year.

TREES, ROADS, WALKS, AND CEMETERIES.

Perhaps the noblest feature of Parisian gardening or Parisian improvements is the great abundance of healthy young trees that are introduced into the very heart of the city, and planted wherever a new road or boulevard is constructed. It is indeed very surprising to see how well this is done, and to what an enormous extent, as well in the centre of Paris, on the boulevards, along the river, &c., as on the scores of miles of suburban boulevards, radiating avenues and roads, the sides of which one would think capable of supplying Paris with building ground for a dozen generations to come. All the planting in all the London parks is as nothing compared to the avenue and boulevard planting in and around Paris. The trees are nearly all young, but very vigorous and promising. Every tree is trained and pruned so as to form a symmetrical straight-ascending head, with a clean stem. Every tree is protected by a slight cast-iron or stick basket, neat wads and ties preventing this from rubbing against the tree injuriously; it is staked when young, and when old if necessary. Most important of all, nearly every tree is fortified with a cast-iron grating six feet wide or so, which effectually prevents the ground from becoming hard about the trees in the most frequented thoroughfares, permits of any attention they may require when young, and of abundance of water being quickly absorbed in summer. The expense for these strong and wide gratings must be something immense, but assuredly the result that will be presented by the trees a few years hence will more than repay for all expense, by the grateful shade and beauty they will afford the town in all its parts. It is almost too much to hope for such improvement in London, though it is capable of being beautified to an even greater extent by judicious planting, and the trees used in Paris would do equally well in it; but we may reasonably hope that trees will be planted along the Thames Embankment; there will be plenty of space and a fine opportunity, and they would highly embellish what will be the greatest improvement London has yet seen.

As soon as a new road or boulevard is made in Paris, in go the trees—and every one of the millions is as carefully trained and protected as a pet tree in an English nobleman's park. The kinds most in use for avenues are the Plane, Chestnut, the large-leaved Elm, the Robinia, and the Ailanthus. Paulownia imperialis is also used here and there; and I noticed that it was in some respects the best tree of all, lasting in the most healthful greenness of leaf long after the common deciduous trees had begun to suffer. It also presents, from its low stature and spreading habit, a great advantage for some places where low trees are indispensable, and shade equally so. Thus, if they planted it in the garden of the Palais Royal, they would have a more agreeable shade than that afforded by Elms, &c.,

while the annual clipping and consequent ugliness of these would be done away with.

As for the making of the roads and streets it is admirable, as many readers may have learnt for themselves. When they have finished repairing or making a street here, the surface is level and crisp as the broad walk in the Regent's Park, so that the horses are spared much pain, and carriage movement greatly facilitated. Stones of much the same size as we spread on the roads are thrown down, and then comes the heavy steam or horse-drawn roller, making but a slight impression at first, as might be expected, and indeed it has to be passed over many times ere the work is completed. But they persevere and roll away day and night, and soon the street looks as described. All the time, or nearly all the time that this rolling is going on, a man stands at the side of the footway in charge of a hose on little wheels, and keeps swishing the stones with water, while others shake a little gravel on them between the rollings, and so they wash and roll and grind day and night, the result being that the Parisian roads are as comfortable for locomotion as could possibly be desired. Doubtless such management as this, the immense improvement effected in Paris gardens of late years, and their perfect keeping, will act beneficially on the horticulture of the country generally; and unless some happy change should start us on the same road to improvement, we shall soon have good reason to be ashamed of the state of our towns and cities. Asphalte is coming into very extensive use in Parisian improvements. Nearly all the fine wide pathways in the boulevards are made with it, and now in many parts noiseless smooth roads are taking the place of those of paving-stones. But, in the case of all the great boulevard and main roads, an improved system of macadamizing is the rule—a considerable breadth on each side being well paved for heavy waggon transit. The asphalte is used in two forms—as a powder and as a liquid. It is used in a liquid state for the making of those fine wide footways, often wider than our roads, that occur so frequently in Paris.

To depose the asphalte properly, the area must be well levelled and very hard; put on the top about one or two inches of *beton* (mortar composed of quicklime, sand, and gravel mixed together), and cover the whole half an inch thick with mortar. If the surface be not a solid one, the thickness of the *béton* should be three or four inches. The surface must be very dry, or the success of the operation may not be perfect.

During the last few years the preparation of the asphalte has been much improved. Some years ago, when a pavement was to be made with asphalte, a great nuisance was experienced by the public during the operation. The matter was liquefied on the spot, and produced a nasty smell and smoke, disagreeable and injurious; but now some of these inconveniences have been done away with by a new system, and asphalte is now laid down in the most expeditious manner. It is prepared first in out-of-the-way places devoted to the purpose, and the matter, ready for use and liquefied, may be transported from these places to any parts of the town without the least inconvenience in a semi-cylindrical boiler, closed by iron doors, and moved about on iron wheels as freely as a common cart. Under the boiler is a fireplace, and the blaze, after having heated the two sides of the boiler, passes out by a chimney placed at the back of the machine. Means to keep the matter in motion, and prevent its burning by adhering to the sides of the boiler, are secured by a simple mechanism easily worked with the hand. These carriage boilers, full of liquid asphalte, are driven from place to place with the greatest facility. The boiler is emptied by the means of a pipe fixed to its bottom, and the matter is collected in a pail, and spread on the surface to the thickness of three-quarters of an inch. If the surface is not perfectly dry, the drying must be accelerated with hot ashes, which are to be taken away afterwards, or with a little spreading of quicklime in powder. These operations are indispensable, as if the asphalte were laid on before the surface is dry, the heat of the asphalte would dispel in steam the water underneath, and that steam would produce blisters in the asphalte, which would crack under the pressure of the feet, and endanger the success of the operation. The operators place on the

platform two iron bars of the same thickness as the asphalte is to
be, equally distanced from each other, and the asphalte is brought in
a very warm state, but thick enough to require some slight exer-
tion of the operator to make it level. This operation done, a small
quantity of fine gravel must be spread upon the asphalte when hot,
and slightly beaten down to penetrate in it. This gives a greater
resistance and solidity to the footway, and insures its lasting for a
very long time.

The roads before spoken of are made of the powdered asphalte.
The surface of the roadway must be beaten down very hard, and
covered at a thickness of about three inches with *béton*, well beaten
down and dry. If the dryness is very necessary in the making of
a pavement, this condition is of a greater importance for the road,
as, if the powder were spread on a wet surface, the steam caused by
the heat would produce a great quantity of little fissures, the elasti-
city would be destroyed, and the road would be useless after a few
months use. The *béton* well dried, the powder (hot) must be
spread about three inches thick; and then proceed with the beating. The sides must be done first, and pressed down with a rec-
tangular iron pestle eight or nine inches in length and two or two
and a half inches in width. When the sides are done, proceed
with the middle. The pestles used in beating are made of cast-
iron, and circular, with about eight inches diameter. The pestles
of either form are heated and used quite hot, so as to compress the
asphalte into a hard smooth mass. When the crust of asphalte is
brought to the thickness required, and is sufficiently smoothed and
beaten hard, they spread with a sieve a little quantity of very fine
powder to fill all the unevenness, and again smooth the whole with
a flat piece of hot iron. The compression is completed by the rolling of two cast-iron rollers, one of 4000 lb. weight and the other of
3000 lb. Sometimes three of these rollers are employed, and the
intermediate one is about from 1500 to 1600 lb. weight. This
rolling is not always necessary, and in many cases the beating
down with pestles is sufficient. The roads thus made, completely
noiseless and lasting a long time, have been adopted with the

greatest success by the city of Paris, and will probably succeed the paving stones, macadamizing, and all the rest of it. Some beautiful smooth roads through the Luxembourg gardens are made of this powdered asphalte, and without the rolling with heavy rollers, the hot "smoothing irons" being used.

Our English cemeteries are often beautiful gardens, quite green, and abounding with trees, weeping and otherwise; while in the country churchyard—

> " There scattered oft, the earliest of the year,
>   By hands unseen are showers of violets found."

And

> " Beneath those rugged elms, that yew tree's shade,
>   Where heaves the turf in many a mouldering heap,"

there is a quiet verdure which makes the spot sweet to look upon; but with the cemeteries of Paris it is very different. There human love is lavish in its testimony, but the result is ghastly to behold. The quantity of the flowers of the sand Gnaphalium that is there woven into wreaths, or immortelles, for placing on and about the tombs in the cemeteries, is something astounding. Next to seeing the contents of a hundred Morgues displayed, the great spread of decaying everlastings is the most ghastly sight. They hang them on the poor little wooden crosses, they pile them inside on the covered tombs, they hang them on the few green bushes, they sling them under little spans of glass placed purposely over many tombs to protect the immortelles from the weather, and in every other way, till in every part, and particularly the part where the second and third class departed are buried, there is scarcely anything to be seen but everlastings in every stage of decay, the *coup d'œil* being most depressing to anybody used to green English or Irish churchyards. A considerable portion of each large Parisian cemetery seems made to be inhabited by ghouls. In addition to decomposing compositæ, there is no end of small crockeryware art, and countless little objects made in beadwork, and brought here by the survivors of the dead to hang on the little black crosses or tombs. It is somewhat different in the portions devoted to the graves of those who could command money when they moved about on the surface, and such as passed on their way to the grave through the paths of fame or glory. In their case, a little chapel, a ponderous tomb, or something of the kind, usually protects for a little time the dust of particular individuals from mingling with the common clay of their poorer relatives, and affords shelter to the crosses of silver and little objects of art, and a little more permanence to the wreaths. But what a very wide difference between this portion and that in which the ground is not paid for in perpetuity! Here the dust is allowed to lie undisturbed (at all events, till they want to make a railway through it, or the gardening taste of a future age directs the surface to be levelled and planted with horrid taste as a garden, as has been recently done in several cases in London), and the earth is not merely a deodorizing medium, as it would appear to be in other divisions. In the select parts, in addition to small statuary, &c., you frequently see choice forced flowers placed on the tombs, and one cold February day I saw a dame, evidently a nurse or respectable servant, sitting weeping by the costly tomb of a young woman buried that day twelvemonth, which tomb she had almost covered with large bunches of white forced Lilac, and beautiful buds of roses. But remove to the wide spaces, where the poorer people bury their dead out of their sight, and you will see a most business-like mode of sepulture. A very wide trench, or fosse, is cut, wide enough to hold two rows of coffins placed across it, and 100 yards long or so. Here they are rapidly stowed in one after another, just as nursery labourers lay in stock " by the heels," only much closer, because there is no earth between the coffins, and wherever the coffins (they are very like soap-boxes, only somewhat less substantial) happen to be short, so that a little space is left between the two rows, those of children are placed in lengthwise between them to economize space ; the whole being done exactly as a natty man would pack together turves or Mushroom spawn bricks. This is the *fosse commune*, or grave of the humbler class of people, who cannot afford to pay for the ground. I am not certain what becomes of the remains of these poor people after the lapse of a short time, but by some means or other the ground is soon prepared for another crop. On this principle, " the rude forefathers of the hamlet sleep " but a very short time in their last bed, and there is a very wide difference indeed between " sickle and crown " in Père la Chaise.

One day, when in the Cemetery of Mont Parnasse, I saw them making a new road, *the bottom being made with broken headstones, many of them bearing the date of* 1860 *and thereabouts.* These had been placed on ground that had not been paid for in perpetuity, and were consequently grubbed up when, as before described, they want to fill the trenches a second time. I have read and admired Lyell's illustration, that " all flesh is grass "—the passage in which he tells us about a graveyard being undermined by the sea on the eastern coast, and a stone inscribed to " perpetuate" the memory of somebody being knocked and ground about by the waves on the beach—but never fully knew what a poor, transient, weedy kind of grass is the flesh of the lords of creation till I became acquainted with Parisian cemeteries. A cutting thirteen or fourteen feet wide, with the earth thrown up in high banks at each side, a priest standing at one part near a slope formed by the slight covering thrown over the buried of that day, and, frequently, a little crowd of mourners and friends, bearing a coffin. They hand it to the man in the bottom of the trench, who packs it beside the others without placing a particle of earth between; the priest says a few words, and sprinkles a few drops of water on the coffin and clay; some of the mourners weep, but are soon moved out of the way by another little crowd, with its dead, and so on till the long and wide trench is full. They do not even take the trouble to throw a little earth against the last coffins put in, but simply put a rough board against them for the night. Those places not paid for in perpetuity are completely cleared off, dug up, and used again after a few years. The wooden crosses, little headstones, and countless ornaments are cleared off, thrown in great heaps, the crosses and consumable parts being, I believe, sent to the hospitals as fuel. The headstones from such a clearing (when not claimed in good time by their owners) go to making the drainage of a drive, or some like purpose. And yet these people, who cannot afford to pay for the ground in perpetuity, go on erecting inscribed headstones, and bringing often their little tokens of love, knowing well that a few years will sweep away these, and that afterwards they cannot even tell where is the dust of those that have been taken from them. What an instance of human love and man's fugacity! Let us hope that whatever else may be "taken from the French," we may never imitate them in their cemetery management.

CHAPTER V.

Floral Decoration of Apartments in Paris.

YOU can grow good plants in England without doubt—nobody denies that merit to English growers; but now-a-days, when the taste for having plants indoors is becoming so prevalent, it is not enough to have good specimens—we should know also how to arrange them tastefully. That in England conservatories are occasionally well arranged is as much as we can say; for the absence of the plants that are indispensable to good arrangement, is too prevalent to permit of that being at all general. And as for indoor decorations in London or elsewhere in Britain—well, of course some nurserymen who "furnish" extensively, can and do arrange plants nicely for balls, &c.; but it is rare to see anything really well done in that line—while the way plants are arranged at the Linnean and Royal Societies and other important places, on special occasions, is almost sufficient to prevent people tolerating plants indoors at all. As compared to the plant decorations of one of the balls at the Hôtel de Ville, anything seen in the British Isles is poor indeed; and yet the plants are not better than may be seen in England in good gardens. The difference is caused by exceedingly tasteful and frequently peculiar arrangement. What the Parisians do as regards arrangement may perhaps be best gleaned if I describe the decorations for one of the balls of the past season at the Hôtel de Ville.

Entering the Salle St. Jean, the eye was immediately attracted

by a charming display of vegetation at one end; while on the right
and immediately in front of and around a large mirrored recess
was a very tasteful and telling display made as follows:—In front
of the large and high mirror stretched forth a bank of moss, common moss underneath, and the surface nicely formed of fresh green
Lycopodium denticulatum, the whole being dotted over with the
variously-tinted Chinese Primulas—a bank of these plants, in fact,
high enough in its back parts to be reflected in the mirror with the
taller plants which surrounded it, gradually falling to the floor,
and gradually merging into the groups of larger plants on either
side of the bank, the whole being enclosed by a low gilt wooden
trellis-work margin.   Then the groups at each side contrasted
most beautifully with this.   Green predominated, but there was
a sufficiency of flower, while beauty of form was fully developed.
In the centre and back parts of these groups were tall specimens
of the common Sugar-cane (Saccharum officinarum) which
held their long and boldly arching leaves well over the group,
and these were supported by Palms, which threw their graceful lines
over the specimen Camellias, these being in their turn graced here
and there by the presence of a Dracæna or Dwarf Palm, and so down
to the front edge, where Cinerarias, forced bulbs, Primulas, and
Ferns finished off the groups, all very closely placed, so that neither
the lower part of the stems, nor a particle of any of the pots, could
be seen, any interstices that happened to remain between the bases
of the plants being compactly filled with fresh green moss, which
was also pressed against the little gilt trellis-work which enclosed
the whole, so that from the uppermost point of the Cane leaves to
the floor nothing was seen but fresh green leaves and graceful
forms, enshrouding the ordinary flowers of our greenhouses, which
are infinitely more attractive when thus set in the verdure of which
Nature is so profuse, and which is always so abundant where her
vegetable beauties are at their highest.

A scene such as this explains the prevalence of these graceful
and noble-leaved plants in Paris gardens and in Parisian flower-
shops and windows, for you may frequently see graceful little

Dracænas ornamenting windows there, and as they look as well at Christmas as at midsummer, I need hardly suggest how highly suited they are for purposes of this kind. The number of Dracænas cultivated in and around Paris is something enormous, and among the newer species of these—not alluding to the coloured-leaved kinds—are some that combine grace with dignity, as no other plants combine them. They are useful for the centres of noble groups of plants in their larger forms, and the smaller species may be advantageously associated with the Maiden-hair Fern and the Cinerarias of the conservatory bench. They are of the greatest utility in these decorations, and are largely used in all parts. So are most kinds of fine-leaved plants, from Phormium to Ficus. So too are young Palms cultivated to an enormous extent about Paris, and every green and gracefully-leaved plant from the Cycads to the common trailing Ivy—used a good deal to make living screens of. With such plants they have but little trouble to find materials for this kind of embellishment. The wide staircase ascending from this *salle* had also a charming array of plants so placed that the visitors seemed to pass through a sort of floral grove—fine-leaved plants arching over but not rising very high, and having a profusion of flowering things among and beneath them. As the bank of Primulas and the groups of tall plants were placed opposite this staircase, and reflected in the great mirror behind, the effect when descending the staircase was fascinating indeed. A still more noble effect was produced in a *salle* near the great dancing saloon, and through which the *invités* passed to the magnificent ball-room. Against each pillar in this saloon was placed a tall palm with high and arching leaves as in Seaforthia elegans, and others with longer leaves and pendulous leaflets. These meeting, or almost meeting across, produced a very graceful and imposing effect, while around them were arranged other plants distinguished either by beauty of leaf or flower, and the groups at each pillar connected by single rows of dwarf plants, closely placed, however, and well mossed in, as in the case of the more important groups. The very close placing of the plants is the peculiar part of the arrangement—you cannot notice any dividing marks or gaps, yet there was no awkward crowding. The fact is that with an abundance of plants distinguished by beauty of form, it is almost impossible to make a mistake in arrangement.

Trellises covered with artificial Ivy, and dotted here and there with artificial flowers, were a useful aid in these decorations; placed behind the groups of plants and on the walls near them, their effect was very good, and of course these Ivy-covered trellises were carried far above the plants—to the top of the walls in fact. So you see they do not spare a little artificial aid, although so well fortified with a supply of fine plants. I saw other evidences of this at La Muette—another name for the establishment before described. A lot of respectable-looking evergreens of pyramidal outline were thrown aside. On closer examination they proved to be composed of boughs cut in a healthy plantation of evergreens, and then taste-fully disposed and tied firmly on a strong stake or small pole, which was plunged in a pot filled with a sort of rough and ready mortar, which soon hardened round the stake and kept it perfectly firm. These were employed during the coldest weather, and doubtless it was thought better to act thus, than to expose valuable shrubs to risk of loss from gas and other evil influences. The fine-foliaged things generally, Palms, Ficuses, and the like, are as remarkable for bearing a great deal of removal, and for not suffering much from the drying atmosphere of rooms, as for their grace and ver-dure. The Camellias were more fully embellished with flowers by the aid of a little management. Of course there were plants in plenty in full bloom at Passy which were not required at the Hôtel de Ville, and from these numbers of flowers were taken off without any stem, thus saving the buds immediately below the bloom. The flowers were then brought into Paris in shallow boxes, a slender wire was slipped through the base of each, turned down a little on the other side, and then pulled back, the flower being next seated upon a couple of leaves on or near the apex of a shoot on one of the plants already in position, and fixed by a twist of the wire. In this way

I

a great number of blooms were added for the night without touching a wood-bud of the specimens upon which they grew.

These arrangements are infinitely varied at the great balls, both public and private; rocks, water grottoes, &c., are occasionally introduced, and very extensive arrangements sometimes made in the open air, in the gardens behind the great houses, &c. The Tuileries Gardens at the time of the great fêtes were largely decorated in this way, each of the numerous lamp-posts having a bed of flowers around it, and the whole scene being turned into a flower garden in a few days. The quantity of flowers required to do this was something enormous, and when it is considered that at the same time great quantities of plants were arranged, both indoors and out, in other great public and private buildings, some faint idea may be formed of the enormous extent to which the plant decoration is carried out in Paris. To go more fully into details would be useless—very few words serve to explain the difference between their and our system of decorating with plants. It simply consists in the use of a far greater number of fine-leaved subjects on their part. This, of course, has a great effect in popularizing the use of plants in houses, for how can you make beautiful arrangements in this way if you ignore the higher beauties of plant form? The fashion as carried out in such instances as the above carries its influences through every grade of society. Thus you see people with a graceful Yucca or young Palm, or New Zealand Flax, in their windows and rooms, who, if in England, would not, in all probability, have had a distinct idea of the existence of such things. The extent to which this taste for floral decorations in the Hôtel de Ville is carried, may be judged from the enormous number of plants grown at Passy for that purpose—the New Zealand Flax which is so very useful for indoor or outdoor decoration being grown to the extent of upwards of 10,000 plants, and Palms and all plants with fine leaves in great quantity. The demand for use in private houses gives rise to a large and special branch of trade in many of the nurseries — one Versailles cultivator annually selling 5000 or 6000 plants of Dracæna terminalis alone.

Some nurserymen cultivate young Palms and fine-foliaged plants generally to an enormous extent, while the trade in forced Roses, white Lilac, and cut flowers generally, is quite a speciality. The only feature of this forcing that we do not practise or do not understand, is the production of the white Lilac seen so abundantly in Paris during the winter and spring; indeed I have seen it in the best condition in early autumn, and in quantity too. To meet with a mass of it in October, quite white and deliciously sweet, is a pleasant surprise to the English visitor. You may see large bunches of it in every little flower-shop as early as the month of January, and it is always associated with the early Violet and the forced Rose. This Lilac is the common kind, and yet it is perfectly white. French florists have tried the white variety, but they do not like it—it pushes weakly and then does not look of so pure a colour as the ordinary lilac one. They force the common form in great quantities in pots, and to a greater extent planted out, as close as they can stand, in pits for cutting. The plants that are intended for forcing are cut around with a spade in September, to induce them to form flower-buds freely, and they commence to force early in the autumn. They at first judiciously introduce them to a cool house, but after a little while give them plenty of heat when once fairly started, in fact, from $25°$ to nearly $40°$ C. $= 77°$ to $104°$ F. At the same time abundant humidity is supplied, both at the root and by means of the syringe, but the chief point is, that from the day they take the plants under glass they are not allowed to receive a gleam of light, the glass being completely covered with the *paillassons,* or neat straw mats, which are much used for covering frames, pits, and all sorts of garden structures in winter. Thus they get the Lilac to push freely, and gather its white blooms before the leaves have had time to show themselves. The great degree of heat—a degree which we never think of giving to anything of the kind in England, and the total shade to which they are subjected, effect the bleaching. The French commence to cut the white Lilac at the end of October, and continue the operation till it comes in flower

in the open ground.   In the same establishments enormous quantities of Roses are forced, small and pretty Roses being in great
demand in Paris through the winter and early spring.

A well-known Parisian cultivator has obliged me with the following article on—

PLANTS USED FOR ROOM AND WINDOW DECORATION IN PARIS.

By M. A. Chantin.

The following are a few notes on the principal plants which serve
for window decoration in Paris.   Among these, the Palms, without doubt, occupy the most important position.   I give below a
list of those which are most generally used, not only because of
their hardy character, but also on account of the very moderate
price at which they can be obtained.

Chamærops humilis and excelsa.   Corypha australis, although
now but little known as a house plant, is destined in a short time to
occupy a foremost place in the decoration of apartments.   It makes
itself conspicuous by its peculiar beauty, and the number of its
leaves, and is, I believe, the most hardy and enduring of all the
Palms for indoor culture.   Cocos coronata and flexuosa are very
elegant, and produce a charming effect.   Latania borbonica is
certainly one of the most *recherché* plants of this family, and is
valued as much for the beautiful green of its leaves as for its
hardiness and elegant appearance.   Phœnix dactylifera, leonensis,
and reclinata are also very much sought after, and are highly
esteemed.   Areca alba, lutescens, and rubra.   The following Palms
could also be used with great advantage in the decoration of apartments; but their high price and great rarity cause them to be not
much known, although they accommodate themselves to the atmosphere of rooms as well as any of those previously mentioned.
Areca sapida, most of the species of the genus Caryota, Chamædorea
amazonica and elatior, Chamærops palmetto, Elais Guineensis, Eu-

terpe edulis, with its finely-serrated and very graceful foliage;
Oreodoxa regia, young plants of which are very frequently used;
Phœnix pumila, P. leonensis, and P. reclinata, Rhapis flabelliformis,
and Thrinax argentea and elegans.

Next in importance to the Palms we must place the Dracænas.
Those which are the most frequently noticed are Dracæna australis,
cannæfolia, congesta, indivisa, indivisa lineata, rubra, stricta, terminalis, and umbraculifera. Those most easily managed, and therefore the most popular for window ornaments, are Dracæna congesta,
rubra, and terminalis. Pandanus utilis, Vandermeerschi, and Java-
nicus variegatus; Cycas revoluta, and the different varieties of
Aspidistra, occupy also a very important place in the decoration of
apartments.

The plants composing the following list, although suitable and
elegant in appearance, are less sought after and cultivated than the
preceding ones, because they are more difficult to manage as window
plants, and require some care and attention. They are more suitable for glazed frames and cases, and may frequently be met with
cultivated in that manner. Several species of Aralia, more especially
Aralia Sieboldi; Bambusa japonica variegata and B. Fortunei variegata; the different varieties of Begonia; most of the Bromeliads;
Caladium odorum, for winter decoration, and the species with
the beautifully-spotted and mottled leaves, for the summer; Carludovica palmata and plicata; Croton pictum, pictum variegatum, and
discolor; Curculigo recurvata, and some species of the genus Dieffen-
bachia. The Ficus elastica is a very elegant plant for a window
ornament, and some years ago was very much employed for that
purpose; but since it has become somewhat common Ficus Chau-
vieri has been substituted for it in many places. There are many
other Ficuses which are suitable for this purpose, and will be found
most useful when they become plentiful enough. Isolepis gracilis;
Maranta zebrina—this is the only species of Maranta suitable for
cultivation in apartments, as all the others speedily succumb to the
hot and dry atmosphere inseparable from a living room. Several
species of Musa are favourites, but principally M. discolor and M.

rosea; Musa ensete is particularly suitable for window culture, but it is still so scarce, and of such a high price, that it is but seldom met with. Pandanus amaryllifolius. Philodendron pertusum was much sought after during the past winter, and has in most places thriven so well that it has given general satisfaction. Several varieties of Beaucarnea are suitable for rooms, and produce a very beautiful and graceful effect when grown in suspended vases or baskets. Rhopala corcovadense; this plant exhales a some-what disagreeable odour, but is nevertheless very much sought after, on account of its very elegant and graceful appearance during the development of its young leaves. Hecktia pitcairni-folia is capital for suspending in baskets, and some of the Bro melias stand well in rooms, and are very useful. Tradescantia discolor, Phormium tenax, Rhododendron, Camellia, Grevillea ro busta, Euonymus, Aucuba, Bonapartea, Agaves, &c., are also fre quent. As flowering plants the varieties of Epiphyllum truncatum are most extensively used.

The family of Ferns, although classed among plants with delicate tissues, and having a great dislike to dry hot atmospheres, never-theless furnishes numerous examples which, with careful manage-ment, add very much to the beauty of apartments. Thus I have very frequently remarked several species of Adiantum, which, wherever they can be preserved in good health, produce without doubt a most ravishing effect. Pteris argyrea, P. cretica albo lineata, and P. serrulata variegata also produce a very fine effect, with their prettily marked foliage. Alsophila australis and Dicksonia ant-arctica are also sometimes employed for decorative purposes in rooms of large dimensions, where their magnificent appearance never fails to produce a very imposing effect. Nephrolepis ex-altata is universally useful, and stands the air of rooms without the slightest injury.

All kinds of plants bearing flowers have paid their tribute to the ornamentation of windows, from the humble mignonette, upon which the patient sempstress loves to turn her weary eyes, to the magnificent orchid that, with its brilliant colours and fantastic forms, fills with grace and beauty the apartments of the affluent.

Until recently, I had little belief in the utility of orchids for this purpose, but experience has shown me that they may be introduced into a drawing-room with perfect success, the plants not having suffered in the least by the change of atmosphere. The most suitable orchids are the various species of Cattleya, Vanda, Aerides, and Cypripedium. Doubtless the time is not far distant when we may venture to try many more kinds than we can now afford to do; but even from what we have already done in that way, I entertain no doubt that the orchid family will eventually furnish the most valuable of all plants for room decoration. True they may not live throughout the year in rooms, as Ficuses and such plants do, but that is not desirable—their appearance, as a rule, not being prepossessing when out of flower. The quality that they do possess, and that makes them so valuable, is, the thick, succulent texture of the flowers generally enables them to continue a long time in bloom in a room. A like kind of texture enables the leaves to stand during the blooming time without injury.

CHAPTER VI.

The Ivy, and its Uses in Parisian Gardens.

THE Irish Ivy is a very old friend of ours : one we have seen beautifying many positions, and one, as we may have thought, sufficiently appreciated and employed. Gaiety and grace I was led to expect in Parisian gardens, but that they should take up our Hibernian friend, so partial to showers and our mossy old walls, and bring him out to such advantage in the neighbourhood of new boulevards and sumptuous architecture, was a very pleasing surprise. That " a rare old plant is the ivy green when it creepeth o'er ruins old," we Britons all know, but that it is no less admirable when mantling with its dark polished green objectionable surfaces in winter, would not appear to have yet sufficiently dawned upon us. Apart from the fact that the Ivy is the best of all evergreen climbers, it is the best of all plants for beautifying the aspect of town and suburban gardens in winter, not to say all gardens. The Parisian gardeners know this fully, and they, taking the Ivy out of the catalogue of things that receive chance culture, or no culture at all, bring it from obscurity and make of it a thing of beauty. To rob the monotonous garden railings of their nakedness and openness, they use it most extensively, and there are parts about Passy where the Ivy, densely covering the railings, makes a beautiful wall of polished green along the fine wide asphalte footways, so that even in the dead of winter it is refreshing to walk along them. And if it does so much for the street, how much more for the garden ? Instead of the inmates of the house gazing from the windows into the street swarming with dust, or splashing with mud, a wall of verdure encloses the garden ; privacy is effectively secured ; the effect of any flowers the garden may contain is much heightened ; and lastly, the heavier rushes of dust are kept out in summer, for so effectively do they cover the railings by planting the Ivy rather thickly, and giving it some rich light soil to grow in, that a perfectly dense screen is formed. Railings that spring from a wall of some height around the larger residences are covered as well as those that almost start from the ground. Frequently the tops of the rails are exposed, and often these are gilt, so that a capital effect is secured. One day, in passing near the Hôtel de Ville, and looking at its traceries, my eye was caught by something more attractive than these : a gilt-topped railing densely covered with Ivy, and between the mass of dark green and the bared spikes at the top a seam of light green foliage, here and there besprinkled with long beautiful racemes of pale purplish flowers. That was the Wistaria, one of the most beautiful of China's daughters, here gracefully throwing her arms round our Hibernian friend, and forming a living picture more pleasing to the eyes of a lover of nature than any carving in stone. If there are tall naked walls near a Parisian house, they are quickly covered with a close carpet of Ivy. Does the margin of the grass around some clump of shrubs or flower beds look a little angular or blotchy ? If so, the Parisian town gardener will get a quantity of nice young plants of Ivy, and make a wide margin with them, which margin he will manage to make look well at all times of the year—in the middle of winter when of a dark hue, or in early summer when shining with the young green leaves. When the Irish Ivy is planted pretty thickly and kept neatly to a breadth of, say, from one foot to twenty inches, it forms a dense mass of the freshest verdure, especially in early summer, and of course all through the winter, in a darker state. The best examples of this edging that I know of anywhere are around the gardens of the Louvre, and in the private garden of the Emperor at the Tuileries. In the private garden of the Emperor the Ivy bands are placed on the gravel walks, or seem to be so ; for a belt of gravel a foot or so in width separates the Ivy from the border proper. The effect of these outside of the masses of gay flowers is excellent. They are the freshest things to look upon in that city, all through the months of May, June, and July. They form a capital setting, so to speak, for the flower borders—the best, indeed, that could be obtained ; while in themselves they possess beauty sufficient to make it worth one's while to grow them for their own sakes. In some geometrical gardens we have panels edged with white stone—an artificial stone very often. These Ivy edgings associate beautifully with such, while they may be used with advantage in any style of garden. A garden pleases in direct proportion to the variety and the life that are in it, and all bands and circles of stone, all un-changeable geometrical patterns, are as much improved by being fringed here and there with Ivy and the like, as are the rocks of a river's bank. It should be observed that an Ivy edging of the breadth of an ordinary edging is not at all so desirable as when its sheet of green is allowed to spread out to a breadth of from twelve to eighteen inches. Then its rich verdure may be seen to full advantage. It must of course be kept within straight lines if the garden be sym-metrical : if it be a natural kind of garden, you may let it have its own wild way to some extent. To fringe a clump of shrubs with it in the English garden, for instance, would answer quite as well. An Ivy border is very easily made. It is better to get a quantity of young plants from a nurseryman, and then plant them rather thickly where the edging is desired. If a wide belt of Ivy is desired, the plants may be put in in two or three rows, as the French do when making such excellent Ivy edgings as are here described. In any case, after the plants are inserted the shoots must be neatly pegged down all in one direction. The reason why Ivy edgings when seen in England look so poorly compared to those in Paris is, that we allow them to grow as they like, and get over-grown, wild, and entangled, whereas the French keep them the desired size by pinching or cutting the little shoots well in, two or even three times every summer, after the edging has once attained size and health.

In nearly every courtyard in Paris the Ivy is tastefully used. I do not think I ever saw the scarlet geranium to greater advantage than in deep long boxes placed against a wall densely covered with Ivy, and that planted also along their front edges, so as to hang down and cover the face of the boxes. Placed thus between two sheets of deepest green, our old friend the scarlet geranium looked particularly happy, and this is only one trifling instance of the capital effect of the Ivy in improving the effect of summer flowers. One of the best known of the floating baths on the Seine has a sort of open air waiting-room immediately outside its entrance—a space made by planks, and communicating with the quay by a gangway. On this space there are seats placed around, on which in summer people may sit and wait for their turn if so disposed, while the whole is elegantly overbowered with Ivy, looking as much at home as if the river was not gurgling rapidly beneath. This was secured by placing deep boxes filled with very rich light soil here and there on the bare space; then planting the Ivy at the ends of each box, devoting the remainder of the space in each box to flowers, keeping the soil well watered, and training the shoots of the Ivy to a neat light trellis overhead. In the garden of the Exposition a pretty circular bower was shown perfectly covered with Ivy, the whole springing from a tub. Imagine an immense green umbrella with the handle inserted in a tub of good soil, boards placed over this tub, so as to make a circular seat of it, and you will understand it in a moment. That and the like could of course be readily made on a roof, wide balcony, or any such position. One sunny early summer day, when the Ivy was in its youthful green, I met with a shallow bower made of it that pleased me very much. It was simply a great erect shell of Ivy not more than five or six feet deep, so that the sun could freshen the inside into as deep a verdure as the outer surface. It may be used with the best taste in the dry air of a room. I once saw it growing inside the window of a wine-shop in an obscure part of Paris, and on going in found it was planted in a rough box against the wall, had crept up it, and was going about apparently as carelessly as if in a wood. If you happen to be in the great court at Versailles, and, requiring guidance, chance to ask a question at a little porter's lodge seen to the left as you go to the gardens, you will be much interested to see what a deep interest the pleasant fat porter and his wife take in Cactuses and such plants, and what a nice collection of them they have gathered together, but more so at the sumptuous sheet of Ivy which hangs over from high above the mantelpiece. It is planted in a box in a deep recess, and tumbles out its abundant tresses almost as richly as if depending from a Kerry rock.

The Ivy is also used to a great extent to make living screens for drawing-rooms and saloons, and often with a very tasteful result. This is usually done by planting it in narrow boxes and training it up wirework trellises, so that with a few of such a living screen may be formed in any desired part of a room in a few minutes. Sometimes it is permanently planted ; and in one instance I saw it beautifully thus used to embellish crystal partitions between large apartments.

CHAPTER VII.

Gladiolus Culture.—Rose-showing.

THE famous old château and forest of Fontainebleau are
interesting to the ordinary visitor, but to the lover of the
Gladiolus Fontainebleau presents an even greater charm
than the most beautiful of those glorious wilds in the forest do to
the landscape artist. And here I may incidentally state that, of
the things to be seen at Fontainebleau, those best worth remember-
ing are far away from the château and even from the garden, which
is to some extent disfigured by those monotonous and ugly lines of
clipped Lime trees so common in France. It is tedious work getting
away from those interminable long straight roads that lead from the
château in every direction; but once in the midst of one of those
wilds where huge rocks and indigenous trees are scattered in about
equal profusion, the visitor will hardly ask himself why Rosa
Bonheur resides in the neighbourhood. However, our theme is
the Gladiolus, and so farewell to the forest.

M. Souchet is the Emperor's gardener, and has been so for
many years; he is also the famous Gladiolus grower, and his own
grounds are quite apart from those of the chateau. He has been
cultivating the Gladiolus for more than thirty years; and it was
cultivated also here by his father. The Gladiolus is the most noble
of our autumnal flower garden ornaments, and one comparatively
neglected by us. There is no flowering plant so well calculated to
improve the aspect of the autumnal garden, of no matter what style,

as this; and M. Souchet is the best grower of it in the world, and has the largest collection.   He grows it in gardens, or rather fields, surrounded by white stone walls.   In fact, his ground was for the most part formerly occupied by market garden cultivators, &c., and these usually surround their gardens by such walls.   He altogether occupies from eight to nine hectares of land with the culture of his favourite, or say about twenty acres English.

The first thing noticeable in this ground is that about half of the land is unoccupied for the current year.   That bare portion is ploughed, and manured, and cultivated throughout the summer as well as in winter, and thus he has fresh land in beautiful condition for his bulbs every year.   Besides, the fact that the ground is bare for a year helps to counteract to some extent the particularly vicious enemies with which he has to contend, as, having no food on the ground for the summer, they are not attracted; and when the ground is rolled between the ploughing and manuring the tracks of the mole cricket are easily seen, and it may be readily destroyed. This idle ground is thoroughly tilled, ploughed, or in some way disturbed six or seven times during the season, and they would like to do it a dozen times if time or labour would permit.   The ground planted this year will of course be empty next, and so on.   Now, over the whole of the extensive piece of ground planted with Gladioli you could not notice a decayed leaf, and all the plants were in the rudest health, some of the varieties growing as much as six feet high.   It was a fine sight at any time of the year to see the magnificent stretch of varied bloom; but the days about the time of my last visit were very hot, and we were obliged to get up very early in the morning to see it at its best.   Although very showy at noon, yet the hot sun had caused the most open flowers to flag a little.   But in the early morning, when the dew hung upon the bloom, and every petal was braced with its freshness, they were fine indeed.   But if fine here, when seen in great quantity unrelieved by a particle of verdure except that of their own pointed leaves, how much better would plants of the same quality be, judiciously associated with other things of beauty, in a graceful

pleasure ground or garden, and pushing up their bold spikes amidst
and near refreshing verdure !

The insect enemies of M. Souchet would prove enough to deter
and defeat most men. He makes ceaseless war against them,
and if they do succeed in destroying a bulb now and then, it
generally forms the guide to their detection and destruction.
Some of us know the mole cricket, but his ravages in England
do not go very far. If allowed his own way for a fortnight in these
grounds, I fear some of the great bulb houses would suffer from
their want of Gladioli in autumn. When this strong and well
armed little brute gets into a bed of choice Gladioli, you cannot well
dig him out as you could if he happened to be in a kitchen garden.
The way he is killed here is so interesting and effective that I must
relate it. M. Souchet explained it to me, but so many "cures"
and dodges for exterminating vermin are not worth the trouble of
trying the second time, that probably I should not have noticed it
had he not called a workman and given me an illustration on the
spot. When the mole cricket goes about, he leaves a little loose
ridge, like the animal after which he is named : and when his
presence is detected in a closely planted bed of Gladiolus at Fontainebleau, they generally press the ground quite smooth with the
foot, so that his track and halting-place may be the more distinctly
seen the next time he moves about. This had been done in the
present instance in the case of a young bed of seedlings. I saw his
track, and a workman who brought with him a rough jar of water
and one of common oil, opened a little hole with his finger above
the spot where the enemy lay. Then he filled it with water twice,
and on the top of the water poured a little oil. The water gradually descended, and with it the oil, which, closing up the breathing
pores of the mischievous thing, caused it to perish of asphyxia, and
in about twenty seconds we had the pleasure of seeing it put forth
its horns from the water, go back a little when it saw us, but again
come forth, to die on the surface, hindered for ever from destroying
valuable bulbs. Being of large size, and very strong and well
armed, even one of these can do a deal of damage in a bed of

Gladiolus, and therefore the moment the workmen of M. Souchet see a trace of the pest they take means to catch it as described, jars of water and oil being always kept at hand. That is only one enemy—*vers blancs* are worse. Of what a vile opponent this is some idea may be formed, when I relate what precautions M. Souchet is obliged to take against it, even for the sake of enjoying a few Rhododendrons. He has built a nice private house near his Gladiolus grounds, and wishing to have a couple of beds of these shrubs within view of the windows near the garden, he has had to build strong cemented walls deep into the earth around each bed, and fill in the bottom with a deep bed of fine sand, so as to guard against the entrance of this dreaded worm into the bed. But it is among his bulbs that most is to be feared. It is simply the larva of the cockchafer, dangerous here in its perfect state also. He employs a great number of people to gather them at the egg-depositing season, has the larvæ picked up after the plough, and one way or another avoids their ravages, though at great cost of time and money.

The soil is a very sandy, not a fluffy one, observe, but one with some holding power, and yet when you get a dry bit of a clod of it, and crumble it fine on a silk glove, you find that most of it sinks through to the palm of your hand, in the form of nearly impalpable sand. Somewhat the same kind of soil occurs in many parts of England, and is always remarkably favourable to the growth of bulbs, alpine, and rare herbaceous plants, &c. Here it is well-manured, and pretty rich and deep, from having been long used as kitchen-garden ground. They prefer horse manure, and that as well rotted as possible. The time of planting is, perhaps, one of the most important things to be acquainted with, and they do it here from April till the early part of June. This late time is not often resorted to, however, though the bulbs may then be planted with safety. They prefer the beginning of May for the general and the safest planting. The medium-sized bulbs give the best flowers as a rule, the biggest bulbs often breaking into several heads. To plant at various times of course will lead to a succession of bloom, another great point in favour of this fine ornament of our gardens. The seedlings flower in their third year. The time of taking up is October, and from the great quantity to be stored this process sometimes goes on to the beginning of November. The plants are mostly in beds, about four feet wide, placed in rows across the bed, from fifteen to eighteen inches apart. The beds are all mulched, and a little alley runs between every two. In very hot weather they are well watered. Each kind is numbered, the scraps of lead on which the numbers are stamped being wrapped round bits of Vine prunings, stuck in the earth. The beds are also carefully examined during the blooming season, so as to destroy all "rogues." Such are the chief points upon which information is wanted—next for a selection of the varieties.

There are altogether between 250 and 300 varieties in cultivation here. The best of the new varieties of the past year, or rather the new varieties ready to send out this year—for one which only flowered for the first time last year will not be ready to sell for years—are Princess Alice, with very large, fully opened flowers, good shape, and charming lilac colour, faintly striped with rose and white—this is of the finest quality; Semiramis, another very large flower, of fine form, rosy carmine in colour, with carmine stripes and a fine white throat; Uranie, a large flower, with a pure white ground, and decidedly striped with bright rose—a distinct and brilliant variety; Bernard de Jussieu, a dark purplish-toned variety, quite new in colour; Stella, a flower with white base and rosy red margins; and E. Scribe, a large flower, of a delicate rose, flamed with carmine red. But a selection of all the other kinds he cultivates is of greater importance, and I give the following, selected by M. Souchet and myself from the several hundred varieties grown by him. We in the first case selected the undermentioned varieties, and then went over them again, marking the very best of all. This second or choicest selection is indicated by an asterisk to all those so chosen :—

Achille.
Anaïs.
Belle Gabrielle.
Charles Dickens.
Cherubini.
*Dr. Lindley.
*Duc de Malakoff.
El Dorado.
Fulton.
Galilee.
*Impératrice Eugénie.
*James Veitch.
*John Waterer.
Lady Franklin.
Laquintinie.
*Le Poussin.
*Le Titiens.
Linné.
*Lord Byron.
*Madame Furtado.
Madame Leséble.
Madame de Sévigné.
*Madame Vilmorin.
Maréchal Vaillant.
*Marie Dumortier.
Mazeppa.

Météore.
*Meyerbeer.
*Milton.
*M. Ad. Brongniart.
*Napoleon III.
Newton.
Ophir.
Oracle.
*Pénélope.
Prince of Wales.
Princess of Wales.
*Princesse Clothilde.
*Princesse Marie de Cambridge.
*Reine Victoria.
Révérend Berkeley.
Roi Léopold.
Rubens.
*Shakspeare.
Sir William Hooker.
Stephenson.
Stuart Low.
Thomas Moore.
*Sir Joseph Paxton.
Vesta.
*Walter Scott.

It is evident there is an ample field from which to select, and a sufficient variety to please the most fastidious. M. Souchet grows almost exclusively for wholesale houses, and a large proportion of the bulbs of these attractive autumnal flowers, which are met with in the stores of the Paris and London nurserymen or seedsmen, are derived from the grounds of this most successful of cultivators.

I cannot close this without acknowledging the great kindness of M. and Madame Souchet, both being as amiable and excellent in private life as M. Souchet is distinguished in horticulture; and some of the pleasantest of the many agreeable visits I have made to great gardens were those paid to M. and Madame Souchet and the forest and gardens of Fontainebleau. On one occasion I was accompanied by my friend Mr. C. Moore, Director of the Botanic Gardens of

Sydney, Australia. He was equally pleased, and remarked that
many of the ravines through which M. Souchet conducted us, in
the remoter parts of the forest, were remarkably like those inhabited
by tree-ferns near the coast, in Australia.

In France the Gladiolus is cultivated much more abundantly than
with us—a state of things which I trust may not long continue, as
nothing can be more worthy of general cultivation, or more calculated to improve the general aspect of our ornamental gardens. To
those intending to employ it, a few remarks on its capabilities may
be useful.

Perhaps one of the best recommendations of this fine bulb is
that its flowers continue to open long after the spike is cut, and
bloom in a vase of water as freely as in the open garden. I have
never seen anything more beautiful or effective than large Sèvres
vases filled with the spikes of the finer kinds in M. Souchet's
house. Many of his varieties grow five feet or more high ;
when cutting them a yard or more of the spike is preserved, and no
arrangement is needed except to insert their bases in the mouth of
the vase, and allow their heads to spread widely forth, placing
a few branches of evergreens, or any verdure at hand, among
the stems, just to give them a little relief. There is no one kind of
flower that could make such a noble combination, and the effect
within the cool, thick-walled French house, on hot days, was of the
highest character. Then we may safely say that the Gladiolus is
the finest of all our flowers for indoor decoration in autumn, its tall
and noble spike entirely preventing it from being used to produce
the dumpling-like effects given by Dahlias and other popular flowers.
But its uses in the open air are even greater ; nothing in the way
of a flowering plant will prove so good for varying and giving a
more graceful and bold character to the flower garden. One
reason probably why it is not oftener well employed with us, is
that it rises above the dwarf materials of which we are so fond. Of
course it may be combined with these with the best taste ; but
there are also many other ways in which it may be used gracefully.

It should be premised, however, that in all cases either a naturally sandy, rich, and deep light soil should be given to it, or one made so artificially. There are many stiff and sticky soils on which it would be much better to avoid its culture, and turn one's attention to things more tolerant of the soil. But the question of soil once settled, let us take the case of a bed of choice Roses in some position near the house. Most probably this bed will present a somewhat disappointing aspect after the Roses are past their best; and even if they continue to flower well, the peeping forth of some splendid spikes of Gladioli here and there will surely not detract from their beauty. Now, to secure this most desirable end, all we have to do is to insert some bulbs of the various kinds of Gladioli in the spaces between the Roses in the early part of May, or thereabouts. Plant them singly here and there, and at about three or four inches deep. Take up the roots in the month of October. Is it necessary to suggest a score of other analogous uses? Need it be said how tastefully they may be introduced just within the edge of the low choice shrubbery, or beds of valuable shrubs on the lawn? Groups of them in the centre of flower-beds would be splendid; and planted thinly here and there among beds of low-growing stuff—say Saponaria, Mignonette, &c.—they would rise above these, and their effect above the surfacing flower would prove very fine indeed. They may be placed in groups or rings around Standard Roses; they will make the most valuable groups in the garden mixed border; and finally, we may make grand beds of them by themselves, or associated with Lilies or Irises. Where they are grown in some quantity it is, of course, best to give them a position on a border in the kitchen or nursery garden, as the formal aspect of anything grown in quantity is not nice for a lawn; but a single bed or two of them might be brought in with fine effect, either in lawn, pleasure ground, or flower garden. It would be well to edge such a bed with some good and bold subject, so as to hide a little the effect of their somewhat lank and naked stems. The Gladiolus has, indeed, many good qualities and uses for our ornamental gardening; but if it merely possessed the qualities described as a plant for indoor decoration, it would prove worth growing for that purpose alone.

SHOWING ROSES IN FRANCE.—A Rose-growing friend has suggested to me that it might be well to mention any novelties in arrangement adopted by the French in showing Roses, but I know of little worthy of recommendation. The great exhibition of Roses at Brie Comte Robert—surprising accounts of which appeared in the daily papers at the time—was, in some respects, a very different affair to what might have been expected from the accounts of it spread abroad. Brie Comte Robert is situated in a very pleasant country, twenty miles or so from Paris—a country without hedges or ditches, yet picturesque and pretty from the number of fruit trees dotted over the land, and with (at the time of my visit) the ears of ripening wheat bending into the straight well-made roads— a country with rich sandy loam and gentle hills, like parts of Kent, but for the main part covered with wide level spreads of wheat and vines. Brie Comte Robert is an ordinary and rather straggling little French town, with an interesting old church traced with the beautiful art of the olden time, and grey with the lichens of a thousand years; and finally, Brie Comte Robert has a *fête* and Rose show, as all the world has been informed. The Rose-show, although pretty and remarkable of its kind, is not quite a marvel, but simply an adjunct of the village fair. Now, the *fête* of a small place like this is not at first sight, or when examined in detail, a thing to be enraptured with. Imagine a grassy yard or small field, in the centre of which are a few tables and the little hut of a person who divines the future; and here, and all around, a lot of small, meagre, clout-covered tents occupied with various things, from temporary restaurants and gingerbread stalls down to diminutive billiards and little games in which the yokels of the district invest a sou a time, and now and then win a trifling work of art worth about a centime. Imagine, in short, the mildest and smallest corner of Donnybrook fair with every drop of " divilmint" squeezed out of it, and you have a pretty good idea of the sight that greeted

my eyes as I entered the show-yard of Brie Comte Robert. But
at one end there was a very large oblong tent, and on entering that
a very different sight presented itself. There all was fragrance and
beautiful colour. All the Roses were placed on the ground—no
stages of any kind being used. First of all, there ran right round
the great oblong tent a sloping bed of sandy earth, about five feet
wide, covered with young Barley, the seed of which had been sown
eight or ten days before. On this were thickly placed the Roses—
eight rows deep, or thereabouts. They were for the greater part
placed in small earthenware bottles, about five inches high, with
long narrow necks and wide globose bases; and, placed amongst
the Barley-grass, these looked very well indeed. Generally three or
more Roses were placed in each bottle, which was made of ordinary
garden-pot stuff, and of the same colour; and they looked so much
better than those of glass used by some exhibitors, that their use
should be made compulsory. Thus the most conspicuous thing in
the tent was a dense bed of Roses around its sides. In the central
parts of the tent there were beds of various shapes in which the Roses
were plunged in moss, and mostly arranged in masses; for example,
a bed of 700 blooms of Général Jacqueminot, edged with a line of
Aimée Vibert; a bed of Madame Boll, edged with white and red
Roses, all the flowers plunged singly in dark green moss, and so on.
The competitors vied rather in quantity than in quality, and one
exhibitor showed as many as six hundred varieties, or supposed
varieties—certainly he had that number of bottles. Others showed
large numbers also, but in most cases the Roses were inferior to
those seen at an English show. As for the varieties, they were
chiefly such as abound in England. There were quantities of that
fine Rose, Maréchal Niel, to be seen, one bed of it being ten feet in
diameter, the blooms plunged singly in moss. The largest exhibitor grouped his flowers very prettily by arranging wavy lines of
yellow and white varieties through the long mass of rose and dark-
coloured ones.

NOTE ON ROSE CULTURE IN POTS. — Visitors to the Paris

flower markets and shops must have frequently remarked the profuse way in which Aimée Vibert and other Roses are flowered in pots; while even many good cultivators in England, who attempt the culture of the same kind, frequently find it running all to stem and leaf, and flowering but very sparsely. The way the French cultivators secure such a profuse bloom, is simply by selecting the buds from stubby flowering shoots. Thus buds taken from comparatively weak, free flowering, and dwarf shoots produce quantities of bloom—those taken from long " water-shoots" produce little besides flowerless and useless wood.

CHAPTER VIII.

The Cordon System of Training Fruit Trees.

 GIVE this precedence not on account of its importance, but because it has been so much spoken of both in the leading journal and in all the gardening papers; Mr. *Punch* even informing the world that the writer of the articles advocating cordon training in the *Times* was to be made a " Grand Cordon of the Legion of Honour!" At first I merely mentioned the system incidentally, but many letters and articles were written about it, and eventually it appeared to assume an importance which it scarcely deserves. To state what it really is in the hands of the best French cultivators, and what it is worth, is my object in this chapter. The first thing we have to settle is, What is a cordon ? There has been some little discussion upon this point—discussion that was utterly needless, and even mischievous, as tending to prevent the public knowing exactly what the term is used for. In France (and in this country since the subject has been so much talked of) it simply means a tree confined to one single stem : that stem being furnished with spurs, or sometimes with little fruiting branches nailed in, as in the case of the peach when trained as a cordon. Some contended that it meant any form of branch closely spurred in; but this is quite erroneous. The term is never applied to any form of trees but the very small and simple stemmed ones. The French have no more need of the word to express a tree trained on the spur system than we have, and they have trained trees on that system for ages without ever so calling them.

Before this name was applied to the forms herein illustrated, or rather before they came into use, it was chiefly applied to a mode of training vines horizontally—each plant resembling the double cordon, except that the stem was longer for those vines that covered the upper portion of wall ; the vines *en cordon* being trained one over another. However, to settle the use of the term, I wrote to Professor Du Breuil, the leading professor of fruit culture in France. His reply was thus alluded to in the *Gardener's Chronicle* :—"What a vast proportion of controversy and dispute might be saved, would people only agree as to the meaning to be attached to words. Just now, as it appears to us, a great deal of unnecessary discussion is raised as to the word 'cordon.' A wrangle about words is about as satisfactory as an argument to prove a negative. It may serve, perhaps, to stop this futile wordy debate to give the opinion of M. Du Breuil himself on the matter. This renowned horticulturist, in a recently written letter, which has been submitted for our inspection, says that he applied the word 'cordon' to trees consisting of a single branch, bearing fruit-spurs only, and thus resembling a rope or cord. When there are two such branches, M. Du Breuil applies the expression 'double cordon.' In order to be quite accurate, we subjoin M. Du Breuil's letter *verbatim et literatim :—*

"*Le mot 'cordon' dérive en français de cord : j'ai employé cette expression pour désigner les formes d'arbres dont la charpente se compose seulement d'une seule branche qui ne porte que des rameaux à fruit. La charpente de ces arbres rassemble alors à une corde ou cordon. Lorsque la charpente de l'arbre se compose de deux branches, je donne à cette forme le nom de 'cordon double.'* "

I should not thus define at length the meaning of the word were it not that any other acceptation of it would not only be contrary to the generally received laws of nomenclature, but mischievous as preventing the proper understanding of the system, or any benefit being derived from it. Professor Du Breuil states distinctly in his book that, struck with the long period it took to cover a wall by means of the larger forms of trees, he invented those quick-rising simple-stemmed kinds to cover the walls rapidly and give an early

return.   Now it is clear that if we call a fan, or horizontally trained
tree, a " cordon," we not only misapply the term, but prevent the
inventor's very clear idea from being understood.   To show how
erroneous is the impression that the term applies to any kind of tree
with the branches closely pinched in, I have merely to state that
the cordon peach trees in French gardens are never pinched in in
this way, but have the wood regularly "nailed in," as shown
by Fig. 33.   However, the following figures will give a correct idea
of what the cordon system is.

The Apple as a Cordon.

Fig. 24 shows the ordinary simple cordon of the French gardens,
and the mode of fixing its support.   A simple galvanized wire is
attached to a strong oak post or bit of iron, so firmly fixed that the
strain of the wire may not disturb it.   The wire is supported at a

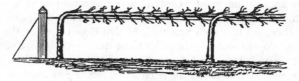

FIG. 24.—The Simple Cordon.

distance of one foot from the ground, and tightened by one of the
handy little implements described elsewhere in this volume.   This
raidisseur will tighten several hundred feet of the wire, which need
not be thicker than strong twine, and of the same sort as that recom·
mended for walls and espaliers.   At intervals a slender support is
placed under the wire in the form of a bit of slender iron with an
eye in it.

The form shown above is used to a great extent in France,
and as I hear from M. A. Leroy of Angers, it is extending with
"extraordinary rapidity."   This and the next are the kinds best
suited for making edgings around the squares in kitchen gardens,
&c.   Cordons are trained against walls, espaliers, and in many ways,

but the most popular form of all, and certainly one of the best and most useful, is the little cordon apple trained to act as an edging to the quarters in the kitchen and fruit garden. Fig. 25 represents the Bilateral Cordon, useful for the same purposes as the simple one, and especially adapted to the bottoms of walls, bare spaces between the fruit trees, the fronts of pits, or any low naked wall with a warm exposure. As in many cases the lower parts of walls in British gardens are quite naked, this form of cordon offers an opportunity for covering them with what will yield a certain and valuable return. It is by this method that the finest-coloured and best apples, sold in Covent-garden and in the Paris fruit shops at such high prices, are grown. Why should we have to buy these from the French at such a high rate? Considering the enormous number of walled gardens there are in this country, there can be no doubt whatever

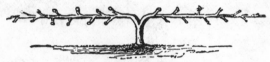

FIG. 25.—The Bilateral Cordon.

that by merely covering, by means of this plan, the lower parts of walls now entirely naked and useless, we could supply half a dozen markets like Covent-garden with the very choice fruit referred to, and be entirely independent of the French. Doubtless many think that these very fine fruit require a warmer climate than we have for them. The French think so too, and therefore place them against the bottoms of their walls. By doing the same we may produce as good or a better result, and may, in addition, grow tender but fine apples, like the Calville Blanche, that do little good when grown as standards. The climate in most parts of England will be found to suit them quite as well as that of Paris, if not better, because the sun in France is in some parts a little too strong for the perfect development of the flesh and flavour of the apple. There is no part of the country in which the low cordon will not be found a most useful addition to the garden—that is, wherever first-rate and handsome dessert fruit is a want.   But in very cold and northern parts, where many apples ripen with difficulty, it will prove a great boon.   There, of course, it would be desirable to give the trees as warm and sunny a position as possible, while the form recommended for walls should be used extensively.   In no case should the system be tried except as a garden one—an improved method of orcharding being what we want for kitchen fruit, and that for the supply of the markets at a cheap rate. The Calville Blanche apple, the kind that is above all others the one I would recommend for growing on the sunny walls of little pits and the naked places above alluded to, sells in Covent-garden at half a crown for each fruit, and sometimes three shillings.

Fig. 26.—Calville Blanche Apple trained as a Cordon.

This is a fair sketch of a specimen of this fine apple I saw growing with others against the bottoms of walls.   The grower told me he hoped to send them to the London market.   So high is the price for the finest specimens of this variety that sometimes the little trees more than twice pay for themselves even during the first year after being planted.   Few but those who know the actual state of the case, would suppose that in this fine apple-growing country we should pay so much for French grown fruit of a variety which we may grow to equally great perfection in the southern parts of England and Ireland.   Yet such is the fact.   I am, however, confident that it will not long be the case.   Of course many other fine apples may be grown in this way, and the increase in their value, even if tested by the best of all tests, the market one, will soon more than compensate the cultivator for the expense incurred by the introduction of this mode of culture. Our own first-class and hardy kinds may be grown as the edgings frequently recommended; the fine but in some cases tender French and American kinds, will be better against low walls, &c., while the fruit borders, with which we have hitherto been in doubt what to do (some contending that they ought not to be cropped at all), will form an excellent position for growing to the greatest perfection our own first-class kinds of apple, any pears that will

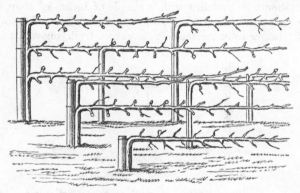

Fig. 27.—Border of Cordons at Versailles.

conform agreeably to the system, and any other fruits that may in time to come be found to do well *en cordon.* Should we find that other kinds of fruit may be grown with advantage in this way, so much the better; but even if we should not, covering the fruit borders of our gardens with the kind that we already know to do well trained thus, will prove one of the neatest and most useful improvements that our fruit gardens have witnessed for many years.

By planting our borders thus we do away with the necessity for disturbing the border after planting, the roots of the wall trees are perfectly safe, a slight extension of the protection usually devoted to the wall will also protect the border, the quantity of noble apples that may be gathered will make it well worth while to contrive an efficient awning for the protection of both border and wall in spring, not an inch of space need be lost, and the border would, when neatly covered with cordons on tightly strained galvanized wires, look well at all seasons.   Instead of planting superimposed cordons, as show by Fig. 27, it would be better to have all single cordons at a foot high, as then the wall would not be in the least shaded; and there can be no doubt that the single low cordon well developed is the best form.   Incidentally I may here state that the borders should be mulched with a couple of inches of short stable manure to allow the roots of all to come near the surface to feed without danger from surface drought caused by drying winds or the

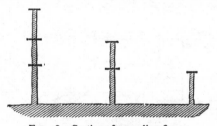

FIG. 28.—Section of preceding figure.

vicissitudes of cold and heat which the surface is liable to.   Good cultivators in France are very partial to covering the ground with a couple of inches of short litter, or mulching as we call it.   Although it has been remarked that it is better to depend on one single line of cordon than adopt the superimposed ones grown in the Imperial garden, and shown at Fig. 27, yet as some may desire this method, and some positions suit it, it may not be amiss to remark that the Doucin stock will generally be found the best where two or three rows are trained one over another.

If such fabulous prices as those above mentioned were never realized, the finer apples that require more heat than they get when fully exposed in our climate would be well worth growing in this manner. Indeed, from an ornamental point of view alone it would be desirable to cover many naked spaces on walls, &c., with these cordons, and embellish them with their pretty flowers in spring and noble fruit in autumn.

Fig. 29 represents quite a young line of Reinettes used as an edging, and, though as yet weak, very fertile. The way of fastening is somewhat simpler than the first cordon figured, and the position of the little *raidisseur* is shown. When lines of cordons are perfectly well furnished the whole line is a thick mass of spurs, just like the top rod of a well-trained espalier. Some keep them very closely pinched in to the rod, but the best I have ever seen were allowed a rather free development of spurs, care being taken that they were regularly and densely produced along the stem. If anybody will reflect that as a rule the best vigour of the ordinary

FIG. 29.—Reinette du Canada as a Cordon Edging.

espalier tree flows to its upper line of branches, he will have no difficulty in seeing at a glance the advantages of the horizontal cordon, particularly if he bears in mind that the system as generally applied to the apple is simply a bringing of one good branch near the earth where it receives more heat, where it causes no injurious shade, and where it may be protected with the greatest efficiency and the least amount of trouble. It is simply a carrying further of the best principles of grafting and pruning—a wise bending of the young tree to the conditions that best suit it in our northern climate. The simple fact that by its means we bring all the fruit and leaves to within ten inches or a foot of the ground, and thereby expose them to an increase of heat, which compensates to a great extent for a bad climate, will surely prove a strong argument in its favour to every intelligent person. I believe it to be the best and soundest of all forms of the cordon system (this opinion is only given after having seen it afford a good result in very many gardens), and that the day will yet come when this fact will be patent to every British gardener. A fruit grower of great experience and repute has objected that it is only fit for small gardens, and ridiculed it by pleasantly describing how the wire tripped one over into the cabbages, &c. Well, if the cordon be no better developed than to be invisible, the less we have to do with it the better; but where it is thickly and regularly set with a stubby spray of fruit buds, and a dense crop of noble fruit, as I have seen it at Ferrières, at Chartres, and at many other places, then it becomes a thing which catches the eye for its beauty and utility. If I were making a garden to-morrow as large as Frogmore, I would run a line of wire round every plot of it at a foot from the ground,

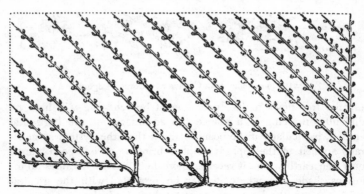

FIG. 30.—The Double Oblique Cordon.

and on that train the best kind of apples, believing this cordon to be much better, more useful, and more easily managed than either the bush or pyramid on the same stock! In numbers of gardens it may be adopted to an extent sufficient to supply the fruit-room with splendid apples without devoting a special quarter to them, or, in fact, losing any space thereby. The wood, leaves, and fruit are more fully exposed to the sun than in the case of either pyramid or bush, or any other method of growing apples away from walls—an advantage for all parts of England, but especially so to cold, northern, and elevated parts.

The form is so definite and so simple, that anybody may attend to it, and direct the energies of the little trees to a perfect end, with much less trouble than is requisite to form a presentable pyramid or bush. It does not, like other forms, shade anything—not even so much so as some vegetables—for beneath the very line of cordons you may have a slight crop. They are less trouble to support than either pyramid or bush; always under the eye for thinning, stopping, &c.; easy of protection, if that be desired; and very cheap in the first instance. Therefore this is the best of all known modes of obtaining first-class garden apples.

It has been objected that ground frosts would prevent the setting of the fruit of these cordons, flowering as they do so near the ground. But the apple flowers late, and usually has a better chance of escaping than the pear; and it is for the apple that the system has been chiefly recommended. I have frequently seen fine crops of fruit from trees that had not been protected. But assuming for a moment that ground frosts would annually destroy the bloom, I may draw attention to the facility with which the cordons may be protected from frost. The spray of any rough evergreens will generally suffice; but I need not say that if it be determined to adopt a more thorough mode of protection, this method is the one to which it can be most readily applied. Where the fruit wall borders are covered with cordons an extension of the protection given to the wall may be made to do thoroughly. Where the thing is tried on a smaller scale how easy to cover the lines with the cheap barless ground vineries, and to remove these to other uses when the fruit is beyond danger!

A few words are necessary as to the best method of planting and managing the apple trained *en cordon,* and planted around the quarters or on borders. In a garden in which particular neatness is desirable it would be better to plant them within whatever edging is employed; but in the rough kitchen and fruit garden they may be used as edgings. The reason why it is desirable to

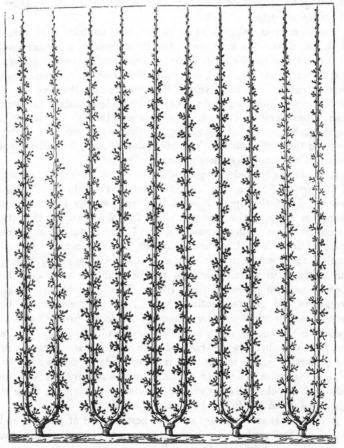

FIG. 31.—The Double Vertical Cordon for Walls, &c.

keep the wire at one foot from the ground is to prevent the fruit from getting soiled by earthy splashings. By having something planted underneath which would prevent this, we might bring the cordon lower down; but though I have thought of several things likely to do this, none of them are very satisfactory. Doubtless, however, we shall yet find something that may be cultivated with profit immediately under the cordon and prevent the splashings, and then be able to bring it within six inches of the earth. Plant at six feet apart or more if the soil be very rich, and such as the apple grows vigorously in. If the soil be a sand, sandy loam, or of a dry nature, use what is called the "English Paradise stock," and if it be a wet or clayey soil use the true French Paradise.

If an opportunity occurs for testing the two stocks side by side, and worked with the same kinds, take advantage of it by all means, and thus ascertain exactly what suits your soil and climate. The French Paradise is invaluable for cold and wet soils, though hitherto it has been condemned as useless "except on those of a very light and dry nature"—the very ones which it dislikes. Test it by all means; it is only by experiments in various soils and situations that we can ascertain the true value of a plant. So far as I have observed, the "English Paradise," though excellent, never gives such very large fruit as the "Pommier de Paradis" of the French. Much care should be exercised as to the selection of kinds. As the system is chiefly valuable for the production of superb dessert fruit, only the finest kinds should be selected. The following will be found very suitable:—Reinette du Canada, Calville Blanche, Cox's Orange Pippin, Duke of Devonshire, Kerry Pippin, Lodgemore Nonpareil, White Nonpareil, Newtown Pippin, The Mother, Early Harvest, Lord Burleigh, and other handsome and well-flavoured eating apples. Wherever it is desired to grow and show splendid specimens of select kitchen kinds, the following will be found excellent:—Beauty of Kent, Bedfordshire Foundling, Lord Suffield, Cox's Pomona, Dumelow's Seedling, Hawthornden, Tower of Glammis, and new or winter Hawthornden, Betty Geeson, and Small's Admirable. There are many other first-rate dessert and cooking apples worthy of culture in this way, and the selection may be regulated by the cultivator's taste and by his soil and climate.

With respect to the form to be attained, little need be said; it is so simple and definite that anybody can manage it. A presentable line of cordons may be made with greater facility than a plantation of bush or pyramidal apples, as all varieties of apples do not conform to these shapes. It is perhaps best to plant the trees some months before bending them to the wire, so that they may be settled down firmly before being attached to the wire. The graft should never be buried. If planted in autumn, it will be best to defer the bending down till spring, as at that season the stems are more elastic from being full of rising sap. When bent down, they should not be severely cut in, or they may betray a tendency to shoot too vigorously from the "bend;" and when the point of the cordon is growing, it may, though tied down occasionally, be permitted to grow upwards a little. By doing so, we encourage the sap to flow on regularly through the stem, and nourish all the spurs on its way; whereas if the growing point be repressed, or too rigorously tied down, there will be a greater tendency to shoot from the part near the bend, and therefore more trouble for the summer pruner; and it is in summer chiefly that the cordons must be pruned. I have seen "cordons" in British gardens with the shoots allowed to rise up like willow-wands. Of course success cannot be attained under such circumstances—such cordons are an eyesore instead of a benefit; and those who cannot attend to them better should content themselves by planting ordinary standard trees. The great point is the summer pinching, to insure a dense array of fruit spurs, while every other suitable means may be used to secure an equal distribution of sap: thus, if the plantation be made on a declivity, all the points should be directed upwards; again, if bare spaces occur on the stem, the dormant eyes should be induced to break by making two incisions a little in front of a dormant eye, and taking out a small piece of bark and young wood. These cares are chiefly required in the establishment of the plantation—once the spurs are regularly developed along the lines, much less trouble will suffice to keep all right. The shoots should be pinched in when they are a few inches long, and to from three to five leaves, according to the strength of the shoot. If very strong, it is not desirable to pinch it in too closely; and although many cultivators pinch in more closely than is here advised, I have certainly seen the best results attained where somewhat freer yet perfectly regular development was allowed to each cordon. And as we confine the trees to one stem, it is the wisest plan not to pinch or repress that too much. The whole is so well exposed to the sun and air, that a dense array of spurs eight or nine inches in diameter will not be too much. Some graft the point of one tree on to the bend of another, and thus eventually make a continuous shoot of each line. This is sometimes so neatly and well done, that the wire may be withdrawn and the trees left to support themselves. When well grafted, and united one to the other, every second stem may be cut away in the event of the plantation becoming too vigorous. Two or three general pinchings during the summer will suffice; but at all times when a " water shoot," or *gourmand*, shows itself above the mass of fruitful little shoots, it should be pinched down. Finally, in winter, the trees will be the better for being looked over with a view to a little pruning here and there; chiefly a thinning and regulating of the spurs when the plantation is thoroughly established, the cutting-in of objectionable stumps, and a firm tying of the shoots along the wire. These should never be tied tightly, so as to prevent their free expansion, but they may be tied firmly without incurring any danger of that. One word more about the pinching. Do not do it when the little shoots are too young; by doing so, a ceaseless pushing forth of soft shoots is the result. In cold and northern parts, where the apple must be grown against walls, the double oblique cordon (Fig. 30) and the double erect cordon (Fig. 31) will be found suitable, especially if enough of fruit cannot be had from the dwarf form which I have recommended for the bottoms of walls, the fronts of pits, &c. The double erect cordon would do very well for running up projecting pillars, or any very narrow space on a wall. However, I introduce them here more to show

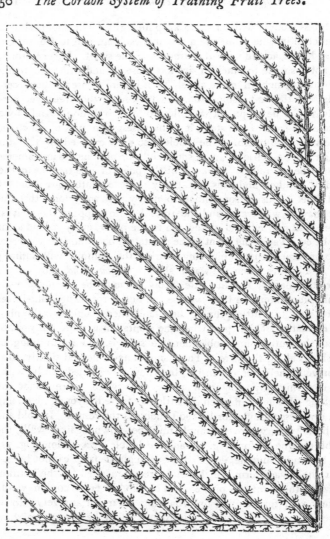

FIG. 34.—Simple Oblique Cordon applied to the Pear.

what the system is than for any merit they possess; they are adaptable and useful in certain positions, but do not possess such marked advantages as the low horizontal cordons.

THE PEAR AS A CORDON.—Having said so much about the apple as a cordon, we will next turn to the pear and the peach. It does not, as applied to either of these fruits, or to other first-rate wall fruits, offer a distinct and economical way of producing a *better* class of fruit, as in the case of the apple. Its advantages are simply quick growth, early fertility, and a considerable number of varieties from a limited space. Here is an example (Fig. 32) of the simple oblique cordon applied to the pear. The plants at each end, which display a fuller development, show the means by which the ends of the wall are covered. As will be seen, the trees are placed very close together, which makes the plantation costly. They, however, soon run up to the top of the wall, and yield a quicker return than the larger forms. Then if one fails it is easily replaced. But are these advantages sufficient to justify us in adopting this system to any extent for our wall pears? I think they certainly are not, and moreover believe we may secure handsomer trees, less distortion, longer life, and more fruit, by adopting such simple and easily conducted forms as those shown at page 159. Those forms are handsomer than the wall or espalier cordon for the pear, yield a great number of kinds from a comparatively small space; and moreover, allow of a somewhat free and natural development. We all know how comparatively few are the varieties of first-class pears which succeed to perfection in any one place, and that the necessity of planting a new kind at every eighteen inches along the wall does not exist. For the fruiting of seedlings and testing of new kinds, it is however a good plan. In these cases it enables us to attain our ends in the shortest space of time, and with the least possible waste of space. The pear may also be trained as a low edging, like the apple, and, with better hopes of success, against the bottoms of walls, the low front walls of houses, &c. As an edging, I have several times noticed it in good condition; but generally the pear pushes a little too vigorously to be trained in that way, while the pendulous habit of the fruit renders it more liable to be soiled. I once saw Uvedale's St. Germain pear grown in this way, the great fruit sitting on the ground, and quite encrusted with earthy splashings. However, some of the best of the free-bearing kinds I have seen used to better purpose as edgings, and know no reason why first-rate varieties of compact fertile habit cannot be grown well in this way, especially on the fruit borders, where they might be protected so efficiently. Planted on a border devoted to peach trees, the same protection might easily be extended to them; and thus we might, by paying a little more attention to securing perfect protection, get splendid crops of both fruits.

THE PEACH AS A CORDON.—With the peach as an oblique cordon, a better result is attained than with the pear, the wall being covered very rapidly, and the neat laying in of a great number of shoots on each side of the simple stem does away with the crowded and unnatural appearance which a plantation of cordon pears assumes when old and the stems are thickened. I have seen some really good results produced in this way, both in glass-houses and against walls.

But instead of the wood being pinched in, as people might suppose in England from reading of the method of one Mr. Grin, it is nailed or tied in at each side of the branch, aye, more so than if that branch were part and parcel of one of the older and larger forms of tree. I once saw an excellent result afforded by this system against the high back wall of a vinery in the establishment of M. Rose Charmeaux, at Thomery. By its means he perfectly covered his wall in a short time, and gathered a great variety of fruit from a small space. Out of doors I have seen it afford beautiful results, and that not unfrequently. It is well calculated for high walls, and it may be adopted for low ones by training the trees at a more acute angle with the earth. Fig. 33 shows how the ends of a wall thus covered with cordons are to be furnished. Considering the time usually

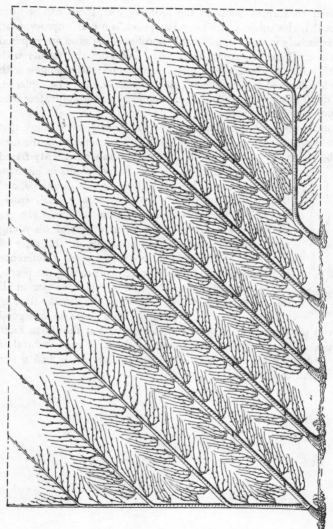

Fig. 33.—Simple Oblique Cordon applied to the Peach.

required to furnish walls in the ordinary way, there can be little doubt that this mode of training the peach is a real improvement, and especially where a considerable number of varieties are required from a small space. Apart from that, however, the facility with which walls may be covered by its means, and the readiness with which a diseased or otherwise objectionable one may be replaced, will doubtless prove sufficient recommendations for cultivators who are not restricted as to space.

I must, however, state that some French fruit-growers think that there is no occasion for resorting to this simple cordon in the case of the peach, any more than in the case of the pear. My friend M. F. Jamain, of Bourg La Reine, plants in his own garden a form of tree with three vertical branches, and if he wants a great variety of fruit from a small space, works a different variety on each branch. Fig. 34 shows, on a small scale, the aspect of one of his young specimens, trained on this principle, at the time of my visit. The aspect of this may suggest a difference in the French and English mode of pruning the peach, and a marked one exists. As a rule they spur-in all the shoots, but not very closely; we lay-in the fruiting wood left after the pruning; and some of our good growers say that the paucity of buds at the base of the shoots must prevent us from adopting the French method. I am doubtful of this, and think that the spurring-in system has never had a fair trial in this country.

FIG. 34.

CHAPTER IX.

Improved Mode of Growing the Pear as an Espalier.

KNOW of no way whereby we may so highly improve the garden culture of the pear as by paying more attention to it as what we call an espalier tree. This is also the opinion of many of the best fruit-growers in Britain, who agree that there is no finer fruit than that gathered from well-managed espalier trees. It is well known that some pears lose quality by being grown against walls. It is equally certain that a fuller degree of sun and exposure than the shoots and fruit get on a pyramid tree is very desirable in many parts of this country, especially for particular kinds. Many grow beautifully as pyramids ; some, to be had in perfection, must be grown upon walls, but by means of the improved espalier system, which I shall presently describe, the majority of the finer kinds may be grown to the highest excellence. If the French can do nothing else they can certainly teach us a lesson as to the improvement in aspect, cheapness, and utility of the espalier mode of growing fruit, and especially pear trees. It should be borne in mind that the good opinion of espalier trees given by British cultivators has been won by the method under the greatest disadvantages, for nothing can be uglier or more inefficient than the usual mode of supporting and training espaliers in our gardens. It is generally costly and disagreeable to the eye, so much so that it has been done away with for this reason alone in many gardens. I know some important ones near London, and indeed in many parts of Britain, where the espalier support is the most unworkmanlike and discreditable affair to be seen in the place. Great rough uprights of wood, which soon rot and wabble out of position, thick and costly bolt-like wire, cumbrous and expensive construction, and, in a word, so many disadvantages as would suffice to prevent the prudent cultivator from attempting anything of the kind. The form or tree used, too, is such that the lower branches become impoverished, and often nearly useless. Now, to form the support for their espalier fruit trees the French have largely adopted a system which is at once cheap, neat, and almost everlasting. Instead of employing ugly and perishable wooden supports they erect uprights of T-iron, and connect these with slender galvanized wire instead of the bolt-like and unmanageable things that we use. These they tighten with the little *raidisseurs* before alluded to, and then there is an end of all trouble. They manage to erect this seven feet to nine feet high for less than a shilling a yard run. I have seen many English arrangements of this kind, supported by beam-like wooden stakes, and tottering so as to be intolerable in a garden. Then, instead of adopting the common form of espalier tree, with horizontal branches, they use more freely trees of which each branch ascends towards the top of the trellis, and thus secure an equable flow of sap through each branch.

But perhaps the accompanying engraving will give a better idea of both trellis and tree than any description. There is no more important matter connected with our fruit culture than this very point, and, therefore, I should be much obliged to all readers, both amateur and professional, who give the subject attention, as I am sure that by doing so they will be led to largely adopt it, and much improve their fruit culture. The finest stores of pears I have ever seen were in gardens with a good length of tree trained in this manner. The form represented here is that adopted in the Imperial Kitchen Garden at Versailles, and is not very common in France, though very good and simple. It is much better than the cordon or single-stemmed pear tree, because a more free and natural development is allowed to the tree, and at the same time the trellis is covered quickly, and a considerable variety of fruit may be obtained from a small space. It is very extensively adopted by M. Hardy, the emperor's gardener, and upon walls as well as on the neat and elegant trellis, of which he has constructed so much. Of course the Palmette Verrier, the fan, or any other form may be trained on these trellises, but decidedly the best are such as combine the ad-

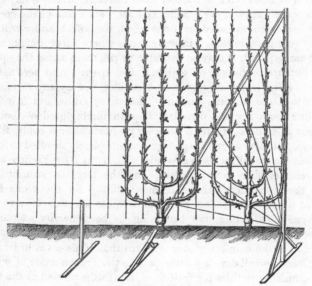

Fig. 35.—Trellis for Pear-growing.

vantages of quick covering and early productiveness claimed for the cordon, and the fuller development and more pleasing appearance of the larger forms. It should be borne in mind that planting erect cordons close together, as they must be planted, involves a great expense which is avoided by using the intermediate forms. It takes a good many years to form the large style of tree generally adopted, and therefore I advise the general planting of these more contracted forms. Nothing can be neater alongside garden walks than lines such as these trained on the trellis alluded to. There is no shaking about of rough irons or wooden beams, no falling down or loosening of the wires; the fruit is firmly attached and safe from gales, the wood is fully exposed, and the trellis thus well covered forms an elegant dividing line in a garden. The best way to place them is at from three to six feet from the edge of the walk, and if in the space between the espalier and the walk a line of the cordons elsewhere recommended be established, the effect and result will prove very good indeed. In some cases where large quantities of fruit are required, it may be desirable to run them across the squares at a distance of fifteen or eighteen feet apart. And here again I mention the Imperial garden at Versailles, not by any means because it is the Imperial garden—it has many faults, and is inferior in not a few ways to a first-class English forcing garden—but the system of trellising there is such as no horticulturist could fail to admire. There is one large square completely devoted to those trellises for espaliers as we would call them (the French apply the term espalier to wall trees); and the foreman informed me that they constructed this excellent trellising for about one franc a yard run or less. The principle is quite simple. However, as the system of wiring now beginning to be generally adopted in French gardens is so excellent and so well calculated to improve and add to the comfort of our gardens that it deserves to be fully described, we will devote a little space to the French mode of wiring walls, making trellises for fruit trees, &c., in the portion of the book devoted to horticultural implements, appliances, &c.

M. Hardy, the head gardener at Versailles, is the son of the celebrated writer on fruit trees of that name, and has had much experience in fruit growing. I here give his written opinion on the trellises and forms of tree herein illustrated. " These trellises are the cheapest as well as the most ornamental that we have yet succeeded in making, and the trees which I plant against them are of that form which I prefer to all others, for promptly furnishing walls and trellises, and to yield a great number of varieties in a comparatively restricted space." The mode of employing the up-

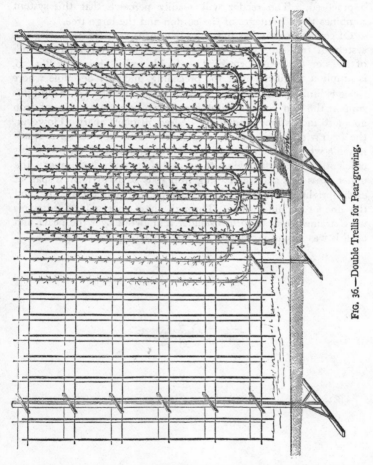

FIG. 36.—Double Trellis for Pear-growing.

rights of pine wood painted green and reaching from the top of the trellis to within six inches of the ground, is not a common one, though very desirable where the erect mode of training the shoots is practised.   The reader will readily perceive that this system combines the advantages of the cordon and the large tree.

Of course many other forms, or any form, may be used with this system of trellising, with slight modifications to suit different kinds of trees or different forms.   The double trellis, shown by Fig. 36, is simply a modification of the preceding, and very desirable where space is limited, and indeed for its economy, for one set of uprights support the two sets of wires simply by using cross bits of iron about 18 inches long, and at the desired distance apart.   However, the engraving shows this at a glance.   This double form is admirable where it is desired to protect the espaliers.   I once saw this efficiently done by allowing the uprights to rise a little above the upper cross-bars and then using iron rods so as to form a span at each upright.   The iron rods were projected about a foot beyond the line of training wires.   Through these rods galvanized wires were strained so as to form a slender framework over the pears, and on this was spread protecting material in spring.

CHAPTER X.

Palmette Verrier.

HEREVER large wall trees are grown, the simple, noble, and beautiful form known to the French as the Palmette Verrier is sure to obtain a place among them. It is indeed the finest of all large forms, and is preferred by many of the best French cultivators to any other. They use it for other trees besides the pear; and by far the finest peach tree I have ever seen was trained after this method near Lyons. The English reader may think it impossible to attain such perfect shape as is shown in Fig. 37, and such perfect equalization of sap as it suggests; but I have seen several trees even more beautifully finished than the one here represented. This figure also shows the advantages of the kind of support used in France for espalier trees as compared with our own ugly method of doing it with rough wooden and iron posts and strong bolt-like expensive wire. It will be seen that the tree differs radically from the usual form of pear tree that we are in the habit of placing against walls, and it is easy to point out its advantages in securing an equal flow of sap to all the branches.

In the common horizontal form the strength and fertility are apt to desert the lower branches, in consequence of their not possessing a growing point to draw the sap through, and particularly when constant care is not taken to repress, by summer pinching, the upper portions of the tree. The form here figured, in common with all very large wall and espalier trees, takes a long time to make. Given a

M

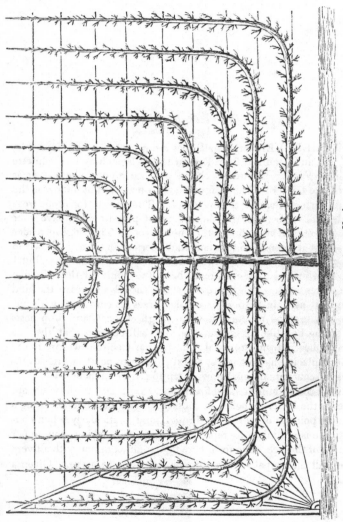

FIG. 37.—The Palmette Verrier.

wall 10 ft. or 12 ft. high, and 20 ft. or 24 ft. long, to be covered with a tree of this shape, it would require fifteen or sixteen years to form it. By adopting a more contracted form based upon the same plan, we may cover the wall or trellis more quickly.

The Palmette Verrier is named after the gardener at l'Ecole Régionale de La Saulsaie, with whom it was first observed. To form the tree, we have in the first instance to plant an ordinary young plant of a desired kind, and of course that should be of the primest quality both as to quality and constitution, as so much care is about to be exercised to make it a handsome and long-lived ornament to the garden and capital aid to the fruit room. It is quite easy to buy trees a little more advanced to make the same form more quickly; but they will cost the more money the further they are advanced beyond what is called the "maiden" stage. Du Breuil, the leading French professor of fruit culture, says the young trees should be allowed to remain a year or so in their positions before being cut, so that they may have rooted well. At the first pruning the young tree is cut down to within a foot or so of the ground, and just above three suitable eyes, one at each side to form the two lowermost branches, the other a little above them and in front to continue the erect axis. Of course all the eyes, except those that are to send forth the three first shoots, must be suppressed in spring. Now, although the tree in the figure looks so very exact and regular in its lines, and the branches appear as if they had been "bent in the way they should go" at a very early stage, it is not so; they are at first allowed to grow almost erect, and afterwards are gradually lowered to the horizontal position. During the first year of the young tree possessing three shoots, care must be taken (as at all times) to secure a perfect equilibrium between them. If one grows stronger than the others, it must be loosened from its position on the wall and lowered. That will divert the sap to strengthen the other one or two. Nothing is more easily conducted than the sap when we pay a little attention to it; if not, it soon rushes towards the higher points, and spoils the tree.

After the fall of the leaf the little trees should have somewhat

the appearance of Fig. 38. We then, at the second pruning, have
to cut them at B, and also cut off about a third of the length of
the side shoots, as at A A, Fig. 38. If one side branch happens to
be stronger than the other, cut the stronger one somewhat shorter.
In cutting and pruning wall trees, the cut should be made above a
front bud, so that the wound made by the knife may be turned
towards the wall, and away from the eye, from which, of course, it
soon will be effectually hidden by this front bud pushing into a
shoot, and thickening at its base. During the second year no
more branches must be permitted to grow, simply because the
trainer desires to throw all the strength he can into the lower
branches, which are to be the longest. Sometimes, however, the

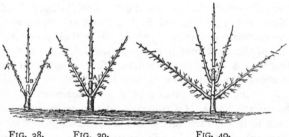

FIG. 38.   FIG. 39.   FIG. 40.

strength of the lower branches will permit the second stage of
branches to be made during the second year of training. At the
third pruning the trees will present somewhat the appearance of
Fig. 39, the central stem being cut at six inches or so above the
previous incision, which is indicated by a slight ring, and a third
part of the *new growth* of the side branches cut off, as shown on
the side branches of Fig. 39. Here, again, we cut above and inside
of three promising eyes to obtain a new stage of branches, and each
succeeding year add another stage; the same thing until the tree is
formed. Fig. 40 represents the aspect of the young tree at the
fourth pruning. At the end of the following growing season the
specimen will have grown sufficiently to allow the lower branches

to be turned up towards the top of the wall, and begin to look shapely. Fig. 41 is an exact representation of what it ought to be at that stage—A, and the cross marks, indicating where the cuttings are to be made. Above all things is it necessary to keep the growth and flow of sap equal, not only for the sake of symmetry, but also to insure perfect health and fertility; for if one part be allowed to grow grossly at the expense of another, an awkward state of things will soon take place. Sometimes, when the vegetation is very vigorous, time is gained in the making of this form by pinching the

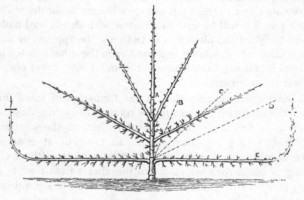

FIG. 41.

central growth at eight inches or so above the highest pair of opposite branches. It then breaks again, and care is taken to secure two side shoots and one erect one. Thus, with care, and in good soil, two stages of branches may be secured in the same year, but this must not be attempted till the proper formation of the two lower branches is secured. The dotted lines in Fig. 41 will show the positions that have been successively occupied by the branch E, when in course of formation, and that it is by no means necessary to train a young branch from the beginning in the exact position it is required to take. In fact, this form is only to be well and easily per-

fected by allowing the young shoots to first grow and gather strength
in an erect or oblique position.   The branch E kept company when
young with the central branch, and was at B ; then it was lowered
to C, next year to D, and finally to its horizontal position.   Some
care is required to make the bend of the shoots equal and easily
rounded.   If the tree be trained on a wire trellis, as depicted in
Fig. 37, it is best to place two bent rods in the exact position
requisite, and before we require the shoot to be bent.   They
must be placed at exactly equal distances from the main stem, and
be equal in curvature.   Then it is an easy matter to gently attach
the growing shoot to them ; it will soon harden to the desired bend.
Against a wall it will be easy to direct it with shreds and nails ; if
the wall be wired the bits of bent twig may be applied, as on the
trellis.   Like care should be bestowed upon the other bends, as they
require to be made ; but of course the outer and lower one is of
the greatest importance.

The reader will observe that, in the formation of this Palmette
Verrier, the custom is not to attempt training the young shoot in the
position it is finally destined to occupy, but, on the contrary, to
permit it first to grow sometimes in an erect, or at least in an
oblique direction, so that the sap may flow upwards without check.
Nothing is easier than taking down the shoots from time to time,
as they become strong and well developed.   Now this is a principle
almost unknown to, and certainly unpractised by, ourselves ; being
applicable to many forms of training, I can strongly recommend
it, having frequently witnessed the good effects produced by carefully carrying it out.   This and the following article are in the
main free translations from the last edition of Professor Du Breuil's
work on fruit trees.

CHAPTER XI.

Pyramidal Training.

AS the amateur reader may wish to know something more of the details of training, and as the French amateur so far excels in the cultivation of pyramidal pears as to have in his garden numerous specimens of which the best English gardener might be proud, I here give the detailed mode of forming these beautiful trees, according to the leading French professor, Du Breuil, in the hope that many British amateurs may be induced to attempt a culture which, while very profitable in all but the more unfavourable and northern parts of these islands, is, considered from the stand-point of beauty alone, as desirable as any with which amateurs interest themselves. I have seen pyramidal pear trees in the gardens of even very humble French amateurs, which, if they never afforded a fruit, were beautiful objects; and I have met with few "avenues" that afforded me more pleasure than a short one of pyramid pear-trees leading up through a little town garden within the walls of Paris.

We will begin, then, with the fully formed pyramid, and in addition to its symmetry will be observed the straight "clean" growth of each branch, springing at regular intervals from the main stem, which is so erect and well furnished.

From the summit to the base, they say, such a tree ought to be garnished with nothing but branches well set with fruit spurs. The greatest breadth of the pyramid should equal about one-third of its

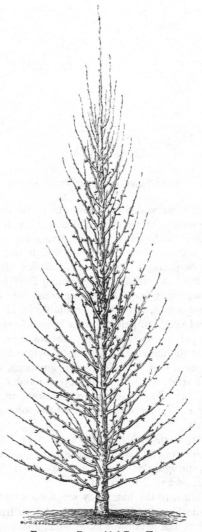

FIG. 42.—Pyramidal Pear Tree.

height. Pyramid trees may be purchased in all stages ; but as trees ready-formed are costly, and as many would prefer conducting their own trees, and those who plant on a large scale will find it econo- mical to begin with trees a year from the graft, we will begin at the beginning with a " maiden tree," letting it grow one year in the ground before pruning it.

Fig. 43 represents the first pruning of this young tree, and its ap- pearance one year after being permanently planted, or two years " from the graft." B shows the union of stock and scion ; and the terminal bud A just below where the shoot is cut should be placed on the side opposite to that on which the scion was inserted, as shown in the figure, so that the erect stem of the tree may rest perpendicularly on its base. It is by attending to such little points as this that they get that perfectly equal distribution of sap which is so essen- tial to the satisfactory management and prolonged fertility of trained fruit trees. The summer following the first pruning the young trees push with great vigour, and their shoots should be thinned when a few inches long, removing every shoot from the base of the stem to a height of about one foot, and thinning out those above this point to six, seven, or eight shoots ; reserving, of course, the best placed

FIG. 43.—Pyramid Pear : First Pruning.

shoots, and taking care to have them arranged as far as possible at regular intervals. Should they in the course of the year assume an irregular development, pinching with the finger and thumb must be resorted to. This is shown by Fig. 44 on the next page. The shoots, A A, have pushed too much ; and one of them rivals the leading shoot B. Therefore they must be pinched, merely taking an inch or so off. In the spring of the follow- ing year the young trees should offer somewhat the aspect of

Fig. 45; the cross marks showing how the pruning is to be performed.

This second pruning has for its object the obtaining of a new set of lateral branches, and the further development of those already

FIG. 44.

FIG. 45.—Pyramid Pear :
Second Pruning.

obtained. It is evident that to secure a beautiful tree, the branches must spring forth regularly from the main stem, which they are not likely to do if the tree is left to itself. Fig. 46 shows the way in which the careful cultivator furnishes his stem, as regularly as could be desired by the most fastidious of pear fanciers. The eyes which he desires to break strongly have an incision made above them, as shown in the figure. This is particularly desirable as regards the lower part of each successive growth of the erect stem ; the vigour of the rising current of sap often pushing towards the higher buds, and causing the lower part to be poorly furnished. These incisions, A, A, A, must be carefully performed on the young branch: deep enough to penetrate the sap wood, and yet not to hurt the slender rising point. But the top of this shoot, instead of being cut off, has been barked for some portion of its length above the bud that has been selected to continue the growth of the coming summer. To this the young shoot is trained, and a perfectly vertical growth for what we may term the pillar of the tree is thereby secured. The bark is

FIG. 46.   FIG. 47.

neatly cut round above the upper eye ; the branch is cut off at about
four or five inches above that point, and then the bark is taken
clean off.   When the young leading shoot is long enough, it is
fastened to the bare portion of stem, as shown at Fig. 47.   The
portion A is cut off at B at the next winter pruning.   This process
may be prolonged as long as necessary or convenient.   In pruning
in the tree considerable judgment is required, so as to get the base
of the specimen well furnished, and secure fertility in the fruiting
branches.   Fig. 50 shows how this is performed, and several of the

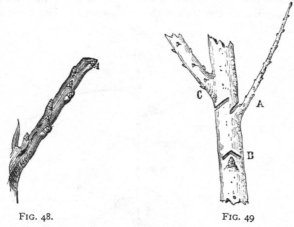

FIG. 48.   FIG. 49

following figures well explain the principle.   It is to cut them of
the greatest length at the base of the tree, and gradually shorten
them as we reach the top.   The nearer they spring to the soil, the
longer they must be left, or to be more precise, only a third must
be cut from the points of the lowest branches; half the length
may be taken from those situated between summit and base; and
lastly, three parts may be cut from the most elevated parts.   In
cutting-in the lateral branches, the directly oblique direction which
it is desirable they should take must be borne in mind in the
pruning, and the terminal bud of each left as far as possible, as

at A, in Fig. 48. In case of a very irregular development among
the laterals, incisions are made above a weak branchlet to encourage
it, as at A, Fig. 49, and below a strong one, as at C, to retard it until
the equilibrium of the branches is established. At B this incision is
made before a dormant bud that has failed to become developed
into a lateral. This figure also shows the relative proportion to
establish in pruning such irregularly developed branches springing
from a main stem that we wish to be equally balanced in all its
parts. The weak shoot is not cut, or but very little; the strong,

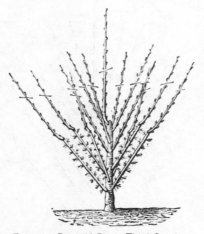

FIG. 50.—Pyramid Pear : Third Pruning.

cut in to below the level of the one it is desired to encourage.
These incisions should be performed with a little saw, so that
the cuts may not soon heal over. The incisions should penetrate
sufficiently into the layer of young wood to well intercept the
sap vessels. If with all these precautions there are objectionably
bare spaces on the stem, they furnish them by grafting by approach
or in other words, turn back a vigorous branch to the main stem,
and graft it on to the bare space ; and if this cannot be done, insert

a short ordinary graft in the stem. That, however, with good management will rarely be necessary.

Having secured the branches straight, the next point is to see that they follow the desired oblique line; and it will be seen by the cuts that the disposition given them is better than what they assume under a less careful system. The light enters freely to the

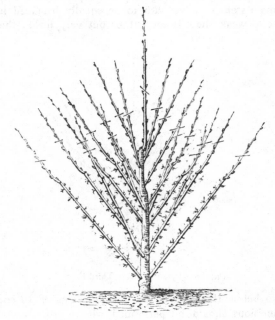

FIG. 51.—Pyramid Pear : Fourth Pruning.

stem, and illuminates all; the more important part of the tree is under command of the eye and hand, and the top is prevented from running away. This, however, is more owing to the fine formation of the lower branches than to the position they assume, though certainly such free and straight outlets for the rising sap are

very effective in preventing a gross development above, and consequently in keeping the tree in the desired condition. During the
summer following the second pruning, the operations for maintaining the lead with the vertical branch, and equality among the lateral

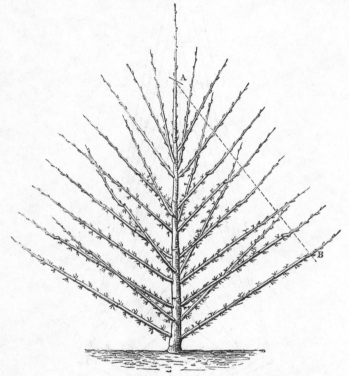

FIG. 52.—Pyramid Pear: Fifth Pruning.

shoots of the new growth, must be carried out as before described.
The third pruning is shown at Fig. 50, and it will be seen that here
again the young lateral branches of the preceding summer are cut

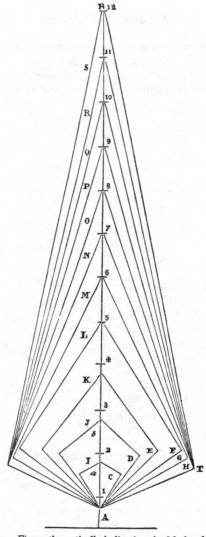

FIG. 53.—Figure theoretically indicating the Mode of Forming
a Pyramidal Pear Tree.

in much shorter than the lower ones to favour the development of these.

At the fourth pruning the lower branches are not cut nearly so long as in the previous pruning, because they have now attained to almost the desired length and sufficient vigour. The new branches of the second series are left somewhat longer, and the pruner looks more to the top structure, so to speak. The wisdom of well forming the base at first will be seen at a glance. During the summer following the fourth pruning, before described, attention should be given to the young branches at the top of the pyramid, while the side ones will also require attention. As the lower branches will have attained to nearly their full length, the terminal shoot of each must be prevented by pinching from a too vigorous growth.

Fig. 52 shows the aspect of the tree at the fifth pruning, and how the pruning is performed. As is well seen by glancing from B to A of Fig. 52, the new growth of the lower branches is cut very short, while the higher the remaining superior branches are, the longer they are cut. A careful look at the figure perfectly explains all this. The succeeding prunings differ nothing in principle from the preceding, future development taking place principally in the middle and higher parts of the tree; and care is taken to train in the correct way by means of twine, and sometimes slender stakes, to guide in the desired direction any branches that may have deviated from it. Thus the pruning is carried on till the tree becomes a large and perfect pyramid, the laterals being well pinched in, and in every case a free terminal shoot is allowed to proceed from each, so that the tree may be kept equally balanced and the sap freely conducted through each branch. They may of course be cut back well every year : always, however, at a likely bud to furnish a shoot for the following season.

Fig. 53 theoretically shows first, the central stem A to B, and its successive cuttings back, 1 to 12 ; secondly, the position successively occupied by the lower branches during the first six years, during which they were successively lowered and elongated from the point C to T ; and thirdly, the lines from I to S show the lines of each year's pruning.

N

CHAPTER XII.

Columnar and Pendulous Training — Pear Growing on Railway Embankments — Suggestions towards Improvement — Short Pinching as applied to the Peach — Peach Culture at Montreuil.

IG. 54 represents a mode of training to be seen here and there in France. The woodcut shows a fully-formed tree before the winter pruning takes place, and, as will be seen at a glance, it is an erect stem densely furnished with short-fruiting branches. This form is considered better than the pyramidal one, where saving of space is a consideration, and where we do not wish the trees to much shade the crops between them. They are also well suited for small gardens where a large number of varieties cannot be afforded space if trained in the usual way. I have thought it worthy of a figure, but except on the quince in suitable soils it is not likely to present many advantages; for if on the pear and confined thus closely to a fastigiate bundle of shoots, it would in all probability run too high to permit of proper annual pruning or of the crop being gathered with convenience. Judging by the strength and thickness displayed by our old horizontal wall trees grafted on the pear stock, what should we arrive at if we adopted a contracted form like this, or the single erect or oblique cordon, with trees worked on the pear? Why, in a few years, and especially with the cordons, we should have objects more like rustic gate-posts than trees.

HORIZONTAL PEAR-TREE WITH BENT
BRANCHLETS.—It is not uncommon in English gardens to train the branches of the
pyramid pear in a pendulous fashion; and it
is a system admired by some, though somewhat more troublesome to form than the
simple pyramid. Fig. 55 represents a mode
of applying a modification of the same principle to the ordinary horizontally trained
pear-tree. I do not say that it is as good as
it is graceful in appearance, believing as I do
in simple easily-conducted forms, but as these
smaller arching branches may be established
on kinds that bear on the young wood, and
with the branches rather thinly placed, it may
prove useful, and is worthy of a trial. The
mode of formation is so simple and easily
established that no further description is
needed. However, I cannot say too often
that the simple and quickly-formed trees,
described elsewhere, are as excellent for
walls as for trellises, combining as they do
the advantages claimed for the cordons with
a not too contracted, repressed development.

FRUIT-GROWING ON RAILWAY EMBANK-
MENTS.—Notwithstanding the scarcity and
high price of fruit in our towns, it is quite
clear that there are many surfaces now utterly useless that would grow it in profusion
and perfection. Many a spot worthless for
ordinary cropping would yield capital fruit
if judiciously planted. There is wall surface now naked which would, if properly
covered with fruit-trees, yield as much as

N 2

FIG. 54.

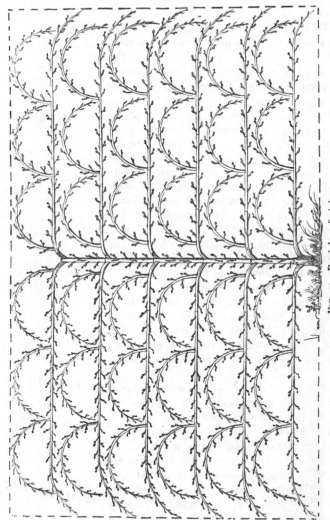

FIG. 55.—Pendulous Training.

would suffice to supply a couple of Covent Gardens; but most conspicuous of all, perhaps, are the railway embankments. Here we have a space quite unused, and on which for hundreds of miles may be planted fruit trees, that will after a few years yield profit, and continue to do so for a long time with but little attention. I am not aware that any attempt has been made to cultivate fruit trees on them in England; but learning that one had been instituted in France, I went one day to see the experiment, which has been made for a distance of eight leagues or so along the line from Gretz to Colommiers—Chemin de fer de l'Est. The French see the great advantage of utilizing spots at present worthless in this way, and are beginning to work at it; but to all intents and purposes they are nearly as backward as ourselves. It is true you now and then hear of somebody becoming a *rentier* by planting a barren mountain side with cherries, but on the whole they have nearly as much to do as we have with regard to fruit culture in waste and profitless places. However, they have commenced, and it is most likely the first trial will be a profitable one, though by no means so inexpensive as like ones might be made. A cheap fence of galvanized wire runs on each side of the line, and on this pear trees are trained so that their branches cross each other, and they are, though only in their third year, nearly at the top of the fence. In some parts they are trained in like manner on the slender but very cheap and slight kind of wooden fence, so common in France. By training them in a way to cross and support each other, before the time the fence decays the trees are perfectly self-supporting, and form a very neat fence themselves. This is a plan well worth adopting in many gardens where neat dividing lines are desired. Some good fruit was gathered last year, but of course the trees are too young to bear anything worth while for another year. Judging from appearances, they will bear abundantly this year, and for many to come. But this, although something in the right direction, does not occupy more than a mere thread of the space on each side of the line, and I cannot but think that much more might be done on the remaining surface by planting small trees.   It would be a great point gained if we could have dwarf productive trees without having to go to expense for fixing or training them—if we could make them self-supporting, in fact.

A well-known French fruit-grower strongly recommends a self-supporting mode of training, of which Fig. 56 is an illustration.   I

have seen examples of it looking quite as well as pears trained on well-made trellises.   M. Baltet says, "It is quite possible and very advantageous to establish neat hedges of pear trees, more or less regularly trained.   By planting them rather close together a quick result is obtained.   At first it would be desirable to train the trees, as

FIG. 56.

shown in the cut, so as to secure a dwarf spreading tendency and equal distribution of sap along the line, but after a time they might be allowed to grow like any common hedge, and even clipped with a shears.   They should be planted at about four feet apart.   The branches should be tied together with loose matting.   If correctly and constantly pruned like espalier trees this arrangement will prove an ornament to any garden, while in its rougher development it may be employed on railway banks as a sort of fence or dividing line.   It resists the strongest winds."

The trees that I saw trained after this fashion were not, however, allowed to get into a rough hedge-like condition, but, on the contrary, trained as neatly and perfectly as ever trees were on trellis or wall.   No flaying of the branches resulted from their being slightly interlaced.   A shoot was taken along the top so as to act as a finish and tend to hold all tighter, and the whole looked much firmer and neater than the ill-supported and ill-trained espaliers that one too often sees at home.   However, the neat and simply-constructed trellises elsewhere described will be much better than this plan for garden use.   The rough modification of it might be used for making the hedges before described; but, on the whole, it seems to me that a mere line of trees, however trained, along a railway, will not effect the improvement we require. Why not plant pyramid or bush trees in such positions? Why not the fig in the southern counties? By covering nearly all the surface of those sunny banks —in many cases of excellent soil—there would be enough work to do to make it necessary and profitable to have men in charge of comparatively short lengths of the line, and these men would be able to better protect the fruit. On the French railway in question the fence of fruit trees is carried along, no matter what the soil or situation. A more rational system would be to adopt the kind of tree to the soil, and simply take the more desirable spots at first.

To sum up the pages chiefly devoted to apple and pear culture, I wish particularly to state that what we must do to improve the stock of our pears, and especially our winter pears, is to pay far more attention to walls and espaliers than we have hitherto done. The cheap walls will prove a great benefit in places where rough material is abundant in the soil; they can be erected almost as cheap as park fencing. The new mode of trellising will also tend much to improve our choice fruit culture, not to speak of amendment in appearances. The adoption of the wide temporary coping will prove of greater benefit than many would suppose : it is more efficient than the far more expensive and troublesome netting coverings. And finally, to devote the fruit borders to fruit culture will be a great advantage. The low cordons will no more shade the wall than a crop of small salading, while preventing all necessity for disturbing the border, and will utilize every inch of its space. Indeed I can conceive of no greater improvement in our fruit culture than devoting those excellent sunny borders that usually lie at the foot of our fruit walls to fruit trees. By doing so we should, it is true, sacrifice some of the more suitable spots for our early vegetables, salads, &c., but we should gain much by settling for ever the somewhat vexed question of the " cropping of fruit borders."

As before remarked, the form of fruit tree of all others best adapted for this work is the low horizontal cordon. The use

of such stretched along a border will not interfere with the
efficiency of a wall in the least degree, while it will be a convenience to have so much of the fruit work together. When the
wall trees are being attended to, the cordons cannot be forgotten,
and the whole will be under the eye at a glance. The pear may
be grown thus, and the apple to the highest degree of perfection;
so much so that I have no doubt whatever that the splendid
apples which may be grown upon this system would, if put to the
market test, more than pay for the expense of protecting cordons
and wall trees at the same time, by an extension of the plan
shown in another chapter. Other fruits will probably be found to
submit to this mode of culture as well as these; and if so, what a
prospect for our fruit gardens! Efficiently protect borders and
walls from the time of flowering till the fruit is beyond all danger,
afterwards expose all to the refreshing summer rains, and then there
would be an end to all but mere routine work till the protecting
season come again. Every one hundred feet length of such well
protected wall and border would be equivalent to the best managed
orchard house; and how very attractive the borders considered
from an ornamental point of view! The fact of the borders being
thus covered with fruit trees will make it almost imperative to
protect the wall and border at the same time; and without efficient
protection at flowering time, we can hope for but very little success
with the finer hardy fruits in this country. The border thus covered
with fruit trees should be mulched, the apple should in all cases be
on the Paradise stock, and the pear, when tried on the horizontal
cordon system, on the Quince, where the soil admits of it. It is
manifest, however, that with the horizontal cordon we may permit
the point of each tree to elongate as much as it seems capable.
By doing that and securing a regular and dense mass of spurs along
each cordon, I see no reason why the pear on its own stock would
not on dry poor soils conform to the horizontal cordon. Protection
from frost while in flower being more necessary with the pear than
the apple, the opportunity of giving it this on the borders proposed
is an additional reason for thus cultivating it.

THE SHORT PINCHING SYSTEM APPLIED TO THE PEACH.—The system generally known among us as that of M. Grin is confounded by some writers with the cordon system, from which it is entirely distinct. This system has been so much written about in England that many suppose it to be in frequent use in France, or at all events accepted as an improvement by good cultivators. This is not the case. It has not in the least influenced the old way of growing the peach in France, and a commission of first-rate fruit-growers sent to examine it, reported that the system pursued at Montreuil is still much the best. It may be shortly described as an attempt to do away with nailing by a system of close pinching, and that alone is sufficient to condemn it for our gardens, and also for those of the French, for the wood to be well ripened must be nailed in, and the pinching required to keep the shoots from running away from the wall is something prodigious. As the French fruit-growers say—the cultivator who pursues this method had better provide himself with chairs, and place one before each tree to accommodate the person who has to see that the pinching is done at the proper time! The report of the commission sent to examine this method is as unfavourable to it as anything can be. I translated it with a view of giving it here, but space prevents my doing so, and therefore I sum up its statements in a few words. " This system, which is an attempt to do away with nailing in of the shoots, presents on the whole no advantages over the one in common use, but, on the contrary, certain drawbacks." Having read so much about the doings of M. Grin, I was astonished at the very ordinary aspect of his trees, and the by no means remarkable result attained. The individual who pays his penny to see the "blue horse captured in the Black Sea by Captain Jones of the ship *Adventurer*—the most extraordinary monster ever seen," &c., in the New Cut, and finds the blue horse to be a puny young seal, could not have been more disappointed than was I at the aspect of the trees in this garden. For when one reads of a system as being about to supplant everything else, it is quite natural to expect that it must at all events possess some merits over the older

one ; but in this instance such is not the case. Of course I speak of
this mode of pinching as a system.

It has one merit, however, and may be used incidentally with
any system of summer-pruning. It should be remarked that M.
Grin commenced by simply adopting a method of very short
pinching-in of the shoot. He now depends chiefly on pinching the
stipulary leaves, as shown at Fig. 57, A. This is the best feature of
the system, and chiefly as respects the little laterals that push forth on
the current year's wood. By pinching the leaves of these little
buds just when they push, as shown in the figure, the development

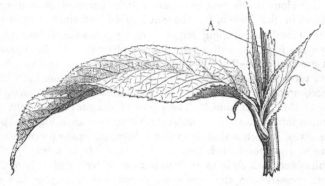

FIG. 57.

of the shoot is not interfered with; but a sufficient check is given
to cause the buds near the base to fill and furnish the shoot as
shown at Fig. 58, A. This not done, the young shoot pushes away,
and is often quite naked of buds at its base. To think of adopting
the system of Grin in its purity would be folly. As to training
peach cordons on this principle, it is simply nonsense; as well might
we think of repressing the flow of the tide as hope to succeed with
trees confined to a single stem, and pinched in quite close. It is
by no means a success even where large forms are adopted; but
with the cordon there is no outlet. The cordons trained after this

method in the public garden at Chartres must have exhausted the patience of the cultivator, for their shoots had started right away from the wall, and grown as much as eighteen inches long!

"This," said I, "will never do for England." "Nor for France!" added an eminent Parisian fruit grower. The only

FIG. 58.

chance of success with the peach as a cordon is by laying in the side shoots regularly. Since the above was written a report on Grin's method has been presented to the Imperial Horticultural Society of France, and in this also a very unfavourable report of the system is given.

PEACH-GROWING AT MONTREUIL.—The limits of this book prevent me from fully describing the accurate and often beautiful way in which the French train the peach, but I hope to do so at a future day. Meantime there are one or two points that must not be passed over. The finest supplies of peaches for the Paris market do not come, as perhaps many would suppose, from the sunny south or the balmy west, but from within a few miles of Paris, where the peach has to be grown on walls furnished with good copings and receive in every way careful protection and culture. Approaching Montreuil the country is seen covered with good crops of vegetables and fruit to the tops of the pretty low hills in the neighbourhood. All the crops, however, are divided into small plots, showing how each person has his own little portion, and has it moreover for ever if he so chooses—land being bought and sold here as simply as an overcoat in England. But getting nearer still to the village, a great number of white walls, about seven or eight feet high are seen, enclosing rather small squares of land, and almost entirely devoted to the peach. As the walls are netted over many acres in some parts, the effect is curious when you look over them from a distance. In the squares are small fruit trees and all sorts of garden crops.

There are numerous cultivators, M. Lepère and M. Chevalier being among the most distinguished. M. Lepère's garden is large, and consists simply of a series of oblong spaces which are surrounded by peach walls, both walls and ground being well covered and cropped—neat, clean, and in all respects satisfactory. Of course the peach is the favourite subject, but neat pyramidal and fruitful pear and cordon apple trees, are also to be seen, and the place is altogether many degrees above the ordinary type of French fruit or kitchen gardens. A very old man, dressed in a blouse, is moving along the walls nailing in the shoots here and there, and with him a dozen young men, his pupils. That is M. Lepère, who has a class twice a week. Incidentally I may say that the principle of giving a full explanation of their system of doing anything well, animates French gardeners more or less. Did anybody ever hear of an unusually successful English market gardener or fruit grower calling a class round him at a low fee, or no fee at all ? The French, though proud of their success in this way, are careful to give it the fullest possible ventilation ; and those who attend here cannot fail to learn the culture of the peach as well as need be, if so disposed, for the master glides along the wall, and stops and nails in the shoots, and cuts out the foremost branches here and there that are not wanted for next year's work ; and, in short, does and explains everything before his pupils. He has been cultivating peaches here for a couple of generations, and certainly has reason to be proud of the result. He inquired as to the state of gardening in

England, and I told him we could beat him in most things, but not with the peach, and that he was indisputably the Emperor of Peach-growers. Entering the garden, your eye for a moment rests upon the perfectly-covered walls, but presently the famous Napoleon Peach presents itself. It is now in good health, but looks a little weak about the central letters. It is, I need scarcely add, beautifully trained, and a striking evidence of what may be done by a good fruit grower—a variety of Adam's race much too rare at present. But looking in another direction a specimen even more striking than "the Napoleon" presents itself, and it takes the form

FIG. 59.

of the letters of the owner's name—LEPERE. It is against a high and very white wall, and at a long distance the letters stand out as clearly as possible, while, upon approaching the tree, the abundance of fruit and regularity of good wood are equally satisfactory. The letters complete, a shoot is taken from the top of each, and these are united in a somewhat arching line above, and spread out again into a great crown above the name, while on each side a single tree springs up, and, forming a border for the letters, spreads out above into a trifid flourish on each side of the crown. It is a finer object than the Napoleon, and bore a splendid crop.

The sketch gives but a very poor idea of the beauty of the tree, which I by no means figure here by way of recommending it or like curious forms, but simply to show the mastery attained over the trees. Such a fanciful form is interesting in a great peach garden, where the grower wishes to show his skill, but is useless for private gardens or for general purposes. It should be added that the formation of the LEPERE was much easier than that of the NAPOLEON tree, inasmuch as a plant is devoted to every letter in the former.

The walls for the greater part run east and west; the soil is of a calcareous nature, and generally the long strips enclosed by the walls are about fourteen yards across. The syringe is rarely or never used, sulphur being the remedy for spider. The ground was in all cases mulched near the trees, a wide alley being left; and for preparation of border they simply trench and manure the ground a couple of feet deep, and about six feet wide. They lay in a good many more shoots than we do, and are doubtless enabled to do so in consequence of the greater amount of sun. The trees are pruned on the spur system, and as for their shapes, they are many, in addition to the alphabetical ones alluded to above. The *Taille en candelabre* is one of the handsomest and most useful. To form it, two branches are taken to the right and left along near the bottom of the wall. From the uppermost single shoots are taken at regular intervals to the top of the wall—the lower branch simply running along to the end, and rising to the top of the wall, or in other words, forming a great oblong frame for the interior. Then there is the *Taille à la Montreuil,* a sort of fan-tail, but with the divisions somewhat far from the base in most cases, and several modifications of the common horizontal mode of training, which we employ so much for the pear, but never for the peach. These seem favourite varieties, and by their means the walls are perfectly covered—if indeed one can draw any distinction between the walls here, which are all as fresh looking as a meadow in May. Generally the tree has a simple erect stem, from which straight branches are sent out, but in an alternate manner, so that the branches of one tree run between those of another. Occasionally, however, two main stems are developed—one to each set of branches. The *Pêcher en lyre*, a very elegant form, is also to be found here, but not to any extent. A form in the shape of a tall narrow U is common. It is simple and readily made. There is also a variety of it which we may call the double U, in which the two branches are again branched, just as they arise from the horizontal, at say about two feet from the centre of the tree; and this too seems a simple and good way of covering a wall with erect branches. The *palmette* and the fan systems are the most appreciated for general work. Much as these trees please when in full health and fruit, to see them when in flower in spring must be much more beautiful.

The reason why the peach is so successfully cultivated here is, it appears to me, that the cultivators pay thorough and constant attention to its wants, with which a life-long experience has made them familiar. They take great pains to have the sap equally distributed, and succeed more perfectly than we do in that important point. The trees are at all times well attended to. I believe that quite as good and as certain results could be attained with the peach in many of the southern parts of England and Ireland, particularly if its culture were made a speciality of as it is in France. When cultivators devote themselves entirely to a subject, they soon learn all its wants, and moreover, attend to them at the right moment— a great point. But it is very different with private gardeners generally, whose hands are so very full of other matters in spring and early summer, a time when the peach requires much attention; and the result is, that it is too often neglected for a week or two at that season, and the result is a loss of health to the trees. There does not seem much help for this in private gardens, and the only hope is that, by the cropping of the borders as elsewhere suggested, gardeners generally may find it worth while to devote more attention to walls than they usually do. I think it a matter for regret that public attention has been to some extent called away from the

many uses and advantages of walls in our climate, and that we have
made no progress in protecting or managing wall trees corresponding with our advances in other respects.   Some persons have gone
so far as to say that garden walls ought to be abolished altogether.
One cannot believe that such people can ever have seen the excellent results produced by well-managed garden walls—results as
beautiful as profitable.   Why, even if we could erect glass-houses
by the economical aid of a magic wand, the good fruit-grower
would still find uses for a large extent of wall surface.   As things
are at present, what all should aim at is greater success in the protection and management of wall trees—a thoroughly practical and
attainable aim.   Our chief want of success at present is due to not
preserving the flowers and tender young leaves from the sleet, cold
rains, and frost, during the cold and changeable spring common to
northern France and the British Isles.

CHAPTER XIII.

Preserving Grapes through Winter and Spring—Preserving the Peach—Fig Culture and Protection.

THE preservation of grapes over the winter with the least amount of trouble is one of the most important of all matters to the British grape grower. Every cultivator, young or old, knows to his cost what a task it is to keep grapes hanging all the winter after they are ripe, especially in places where there are a good many houses devoted to vines. The latest books on the vine give directions for regulating the houses, so as to preserve the grapes on the vine after they are ripe, and every calendar of operations tells how to manage them in that respect, though I fear the directions are not always intelligible.

Here, for instance, is an extract from a recent issue of a leading garden paper :—

"Those who wish to keep grapes hanging as fresh and plump as possible to the longest possible period, must take care not to afford them too much heat, as an excess of this, no matter how dry the structure may be, or how favourably treated otherwise, is sure to cause them to shrivel more or less prematurely. Give only just such warmth to the pipes or flues as will insure sufficient buoyancy to any humidity (!) which may arise in the house as to enable it to make its escape. Independently of the ill effects caused by actual heat, a too warm atmosphere, even in the driest·house, will cause a correspondingly excessive evaporation and consequent condensation."

o

Then of course we must have fire heat and give air when foggy days occur, " as," says Mr. Thompson, of Chiswick—

" The mean temperature of this month (November) is on the average little above 40°, and the air is generally saturated with moisture. When this is the case, moisture will be deposited on all substances exposed to the air, if they are not warmer than it is. Grapes that are ripe should, therefore, be kept warmer than the air, otherwise they will be liable to damp. The application of fire heat would effect this : but if it were applied suddenly, and without air being given at the same time, the heated air would deposit moisture on the berries; for although these would ultimately acquire the same temperature as that of the air surrounding them, yet for a time they would be colder, and so long as this is the case they would act as condensers of the moisture in the warmer air in contact with them. The more rapidly the air is heated, the greater for a time will be the difference between the temperature of the fruit and that of the air, and, of course, the slower the heating the less at any time will be the difference. Give therefore, in damp weather, a little fire heat in the morning and admit air. If the nights are cold, the temperature of the house should not be allowed to fall lower than 45°."

Here, then, are nice operations and a lot of trouble to bestow on perhaps half a dozen houses during the winter months! If the greenhouses are shaky and badly heated. the task is most difficult and annoying; in the best-constructed vineries it is a great, and, as I hope to show, a needless labour. The trouble of regulating the atmosphere, the expense for fire heat, and the necessity of keeping the house almost entirely devoted to the grapes, must render any improvement very acceptable, and I have a decided one to describe. Several times during the spring of 1867 I noticed grapes hanging from branches the ends of which were inserted in vases of water— grapes which the exhibitors described as having been for a long time so preserved in a fresh state. From such few specimens I did not derive sufficient confidence in the method to speak with certainty of its merits, but having visited a good many gardens during the past autumn, just as the grapes were ripe, I found that the method was accepted as a great boon by some of the best gardeners in France, and their system of keeping grapes has been altered accordingly.

The best example was in the gardens of Ferrières, the magnificent country seat of Baron Rothschild. Here they have constructed, in addition to very fine and well-filled fruit rooms, a grape room, which is filled with stands thickly hung from top to bottom with all kinds of grapes—those for present consumption as well as those for use six months hence. M. Bergman, the head gardener, was cutting down all his grapes in harvest fashion, and would in a few weeks, as soon as the latest houses were ripe, have his many and

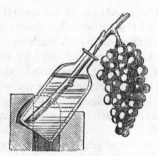 

FIG. 60.   FIG. 61.

well-managed vineries to do as he pleased with : ripen the wood, prune and clean the vines, or utilize the cleared space of the houses for any purpose that might be convenient, not fearing as we do to spill a drop of water or make any other use of the house. The grapes are cut with a considerable portion of the shoot attached, much as if one were pruning the vine ; the shoot is inserted in a narrow-necked and small bottle containing water, and these little bottles are fixed firmly along, so that the bunches hang just clear of each other. In the first instance are erected two uprights, pretty strong, and each supported on three legs. Then from one to the other of these, on both sides, and in an alternate manner, are nailed sets of strong laths, two for each line of bottles. These laths are kept an inch and a half or so apart by a bit of wood at each end; in the inner one there are incisions made, into which the bottom of each little bottle fits, and then the outer lath has a concave incision in which the side of the bottle rests, so that, caught in the inner and leaning firmly on the outer lath, it holds the stem and stout bunch quite firmly. I thus particularize it from having seen other ways of doing the same thing less neat and simple than this. Walking space was left between each wall of grapes; for six or seven rows were arranged one above another on both sides of each support. Fig. 60 shows the insertion of the shoot and fixing of the bottle, and Fig. 61 the structure of the upright.

Charcoal is mixed with water, allowed to stand for some time, and then the water is strained off to fill the bottles. But there can be no doubt that to put a pinch of animal charcoal in each bottle would prove a better plan of guarding the water from any impurity from the slight deposit of organic matter that might be expected; at least, it does not seem very clear how charcoal removed from the water before the vine-stem is put in can have much effect in keeping it pure. However, that is not an important matter, and it is certain that a pinch of animal charcoal, which is very cheap, will keep the water quite sweet. One cultivator who keeps grapes on a large scale by this method, never uses any charcoal at all, but simply fills his little bottles almost full with water, and then inserts the branches, which nearly close the necks of them. He appeared quite as well satisfied with the plan as those who had taken more pains to keep the water sweet. That once settled, all there is to do is to add a little water, in case evaporation should cause that in the bottles to fall below the bases of the shoots. Of course it will be understood in a moment that with one-tenth the amount of expense and trouble that is now necessary in large grape-growing places, we may in a grape-room like this maintain conditions infinitely better calculated for the preservation of the fruit than the atmosphere of any vinery can possibly be. We may keep the fruit dark, preserve the necessary amount of dryness in the atmosphere, keep up a temperature constantly equal—all of which are essential to the well-being of fruits, and none of which can possibly be attained in the house in which the grapes are grown. It would, of course, be wise, in arranging a room of the sort, to have hollow walls and other contrivances to attain the conditions under which fruit is known to keep best. M. Charmeaux, the great grape grower of Thoméry, was, I believe, the first to try it extensively. Now as we grow by far the best and largest quantity of hothouse grapes of any country, this method will prove of far more use to us than to the French. I was told by experienced French gardeners who have adopted the system, that they keep the fruit as long this way as upon the vine, with fewer mouldy berries, and almost without trouble; and it is not likely that a man would cut down half a dozen houses of fine grapes at the beginning of October unless he had already proved it to be a good system.

The advantage of having all the stock of grapes safely housed and away from the attacks of vermin and other interlopers, is another of many advantages presented by this plan, which I now leave in the reader's hands for trial, confident that it will prove a great boon to the grape grower, and tend to make that fruit—every day growing in popularity—a great deal more enjoyable and obtainable in the winter and early spring months. For if it be a process requiring much care in large well-conducted gardens, how much more difficult must it be for the large class of amateurs and small gardeners to preserve their fruit in good condition? In places where the stock of grapes is not sufficient to require a special room for their keeping, part of the fruit room might be adopted, or even a dry cellar or store room.

The above was written previous to visiting M. Rose Charmeaux, with whom the system originated. I have since seen his grapes stored for the winter; the method was in full working order, and even more simple and effective than could have been supposed. He began by having a stove and a couple of chimneys to keep right the atmosphere of his large grape-room; but finding that the grapes keep very much better without this, he now simply devotes to his winter stock a large room in his house, fitting it up in all parts to accommodate handily the little bottles before spoken of padding the inside of the windows so as to exclude light, and obviate, as far as possible, changes of temperature. The grapes are cut in October, and preserved in good condition until April, when his earliest grapes are ripe. He has frequently shown them in May, and even later, and has kept them till August; but of course the quality cannot be expected to be good after such very long keeping, which is merely done for the sake of show. The first result of the method to the village of Thoméry, which is almost wholly occupied with Chasselas culture, was a gain of from 100,000 to 150,000 francs per annum. The system enables the cultivators to keep their grapes much later than of old, and thus to add considerably to their revenue. A small room in M. Charmeaux's house illustrates to a nicety the fact that a similar one in most houses may be made to answer the purposes of keeping grapes. It has no windows, and scarcely any means of ventilation. The rest of the house is heated by hot air; but while there are openings in the floor of the passages and rooms generally to admit this, there are none in the room devoted to grapes. Thus it is clear that the ordinary temperature of a well-constructed house will present suitable conditions for the long preservation of grapes, in any small room that may be devoted to that purpose. The system was attractive enough when it was considered necessary to construct a room specially to carry it out; it is much more so now when it has been proven that not only is it not necessary to take any special means to warm or ventilate the structure, but that the grapes keep very much better without such being taken.

At a recent meeting of the Royal Horticultural Society, Mr. Whittaker, of Crewe Hall Gardens, exhibited a bunch of grapes kept somewhat after this fashion; but as he had not seen the plan in execution, nor the figure of it given in this book, the experiment was not carried out so simply as it may be. However, the grapes had been kept a considerable time, and there was no reason to suppose that their flavour was inferior in any noticeable degree. Notwithstanding this, and that the test was not carried out in the most efficient way, two directors of the Royal Horticultural Society, the Rev. M. J. Berkeley and Dr. Hogg, are reported to have rather emphatically condemned the system. Having, from a very early age, had considerable experience of the keeping of late grapes, I had no more doubt from the moment that I saw this mode of preserving them well put in practice, that it would prove an aid to the British gardener, and be adopted extensively in this country, than I had that the hawthorn will bloom in spring. I therefore took no serious notice of several small attacks upon it by persons who had had no means of judging of it, and little or no influence to prevent gardeners from reaping advantage by its use; but when directors of the Royal Horticultural Society condemn a method that is certain to prove a real improvement and useful aid to the grape-grower, then it behoves me to speak. Prejudice or a hasty judgment is excusable in many cases, but that the leaders of this learned body should condemn a system without having given it a fair trial, or causing it to be fairly tried in their gardens, seems somewhat contrary to what is called the "spirit of the age." The Rev. M. J. Berkeley is reported to have said that the grapes would "lose their sugar;" and if I am not misinformed this opinion first came from Dr. Hogg; but it is probably not an opinion arrived at from experiment in this country, being well known to the French, who are also aware that pears rot after a certain period in the fruit-room. Yet I presume these gentlemen would not argue against storing apples and pears in consequence of the virtues these fruit lose before they become spongy or decayed. The fact is that the French in carrying out their experiments have kept some of their grapes as long as they could for curiosity sake, if nothing else, and have frequently shown them in a nice plump condition long after they ripen their early grapes—just for the honour of the thing. In these instances a loss of sugar was, no doubt, perceptible; but what kind of flavour would berries possess if left hanging on the vine till the summer months when the Frenchmen exhibited their grapes?

The necessity for keeping the grapes till they lose their sugar does not exist. In most of our large gardens grapes are forced early, and would be ripe before the fruit of the previous year had lost its virtues in the least degree. And in our comparatively small gardens, containing perhaps a vinery or two, how many bunches of grapes are left after the consumption of the winter months ?

To be able to clear the vineries of grapes for two months before the ordinary time would be a decided gain to thousands of gardeners in this country. Of course the practical grape-grower knows this well, but as the gentlemen above-named have so decidedly opposed the method, I make no apology for adducing fuller and more minute proofs that we are badly in want of a better system of keeping grapes, and that the method I advocate is thoroughly sound. " About the 15th of April," says Mr. Thompson, " the sap began to rise in the vines, and some of the berries that were a little shrivelled suddenly got plump, while others that had shown no signs of shrivelling burst their skins, and the sap of the vine that had forced itself into them began to drip from them !" Surely in such a case as this it would be a gain to the grape-grower to cut his grapes a few weeks before any danger of such a case existed, and thereby keep them a little longer from bursting their skins and giving forth what cannot be rich in sugar ! The expense and care required to keep grapes during the dull and cold months of winter in the ordinary way is very considerable, and the inconvenience and loss of space great. Here is an extract from " A Practical Treatise on the Grape Vine"—the latest book on the subject. After giving directions as to the heating of the house, its ventilation, &c., and how to exclude foggy damp air (no easy matter in a glass house, by the way), the writer says, " The surface of the internal border is allowed to get perfectly dry, and to remain so all the winter, care being taken that as little sweeping or raking take place as possible, for by this means dust is raised which settles on the bunches." Practically speaking, houses treated in this way are nearly useless for anything except keeping the grapes, consuming fire heat, and wasting labour. Remove the necessity of keeping grapes on the vines long after they are thoroughly ripe, and the houses will be ready to be filled with plants in every nook at a

season when every inch of glass is valuable. The work just named
relates how the author cut a lot of bunches of grapes in February,
and with them a portion of the stem, which he pointed and inserted into roots of mangold-wurzel—a sufficient proof, even if none
other existed, that something in this way would be a boon. However, there is no person who knows what it is to grow grapes but
is aware that some improvement is desirable.

The Rev. Mr. Berkeley is reported to have said with regard to
a system which is certain to avert much of the expense and inconvenience above alluded to, that it had justly been remarked that
they (the grapes) were certainly not improved in quality, that even
with the addition of charcoal the water might get into a putrid
state, and finally that the method was " universally acknowledged
not to be a success!" With regard to the first statement, nobody
has ever said that the grapes were improved in quality ; but that they
are not deteriorated in any appreciable degree by keeping as long
as is necessary, is well known to those who have tasted grapes
kept thus. With regard to the second statement, I can simply
assure the reader that no putrefaction was observable in the water
thus treated, and add my conviction that a pinch of animal charcoal
will keep the water quite safe, and also that it is very likely the
system will be useful without the use of charcoal at all, especially
in cases when we do not require to keep the grapes for a long time.
I have recently had a letter from Mr. Whittaker in which he states
that the water kept perfectly sweet with him, and that he believes
this mode of keeping grapes to be good, and that it will yet be
universally adopted in this country. As to the third statement,
that the method was " universally acknowledged not to be a
success." I should like to ask, by whom? By those in England
who know nothing about it and have probably never seen it carried
out? or by some in France who judged of it from specimens
kept a very long time? M. Rose Charmeaux and other cultivators at Thoméry—the place of all others to test its merits—
are, though the best calculated to judge of its merits, certainly not
among those who have condemned it ; for when I was there last

autumn the large rooms in their dwelling-houses were specially devoted to keeping grapes thus, and hung from floor to ceiling with
well-flavoured fruit. That Baron Rothschild, who has in Seine et
Marne one of the finest gardens and country-seats in existence, with
every means of growing and keeping grapes in the ordinary way,
should have a special apartment constructed in which to carry out
this method, and cut down in autumn all the fruit of his late vineries
to be preserved in this way and used during the winter by crowds
of distinguished guests, does not seem a very convincing proof that
the method was "universally acknowledged not to be a success."
But these are facts which can be testified to by Mr. James Barnes,
of Bicton, as well as myself; and if he be not able to judge of its
merits, who is? If I mistake not he saw the October clearance of
the vineries at Ferrières with some envy as well as satisfaction, re-
membering the long period which has hitherto elapsed before the
gardener can do as he likes with his vines or with the contents of
his vineries.

P.S. (April 24th).—Shortly after writing the above I addressed
a note to the *Gardener's Chronicle,* asking the Reverend M. J.
Berkeley for fuller reasons for condemning this mode of keeping
grapes than he had given when addressing the Royal Horticultural
Society, and for some proof that the method had been universally
admitted not to be a success. In his reply in that journal, he takes
no notice whatever of the changes that are well known to occur in
grapes kept for months after they are ripe in the atmosphere of a
glass house, so exceedingly liable to vicissitudes; and, contrary to
his wont, he does not even take any notice of the luxuriant and
interesting fungoid growths which are frequently to be observed
upon grapes kept in the ordinary way; but he goes "theoretically" into suppositions of what changes might take place in a
case which, though so easy to be tested, he has not tried. He
admits, what he before doubted, that the charcoal would keep the
water sweet; and he alludes to the "French method of keeping
grapes in bottles of water hermetically sealed" (!) thus conclusively
showing that he knows nothing whatever about a method which

he has been so ready to condemn. The French do not seal the bottles at all, nor have they ever done so, but simply do as I have described.

PRESERVING THE PEACH IN A FRESH STATE AFTER IT IS RIPE. —It is at the latest moment that I add this paragraph, and do it with some reluctance, not having seen the experiment to which it refers carried out, nor indeed had any means of inquiring fully into the matter. At first I thought it better to have the matter carefully tested by several people next autumn before speaking of it; but here it is, and the reader can take it for what it is worth. The greater number of trials made with it, the more likely are we to quickly know its value. Travelling one day by the Lyons railway from Paris, a French gentleman who had been shooting entered the carriage with his gun and game-bag, and we soon entered into conversation about the country, especially agriculture and horticulture, and finally chanced to speak of the method of preserving grapes above described—a method which he was well acquainted with and quite approved of. He then told me of a new method of preserving the peach in a fresh condition a considerable time after it was ripe, and that it simply consisted of packing the fruit, gathered while ripe before being quite soft, in bran in rough boxes, and placing them in cellar, store-room, or any similar place. He stated that by this means peaches had been preserved in a perfectly fresh and well-flavoured condition many weeks after the latest peaches had been gathered from the trees, and also that some which had been presented to the Emperor some time about Christmas were mistaken by him for early forced fruit. The train stopped again, and my informant bade me good day before I had time to ask him for fuller information as to the extreme time which the fruit had kept well, and other points. That is all I know about the matter. Possibly he may have been misinformed or misled, but he spoke quite confidently of the method, and as if quite familiar with it. As prolonging the season of the peach would be a great gain, the mode is worthy of a trial, which should be made with the fruit

in several slightly different stages of ripeness, and witn several
varieties. If by its means we can succeed in keeping the most
delicious of all our fruits a few weeks longer than at present, I shall
have reason to remember with pleasure my few minutes' chance
conversation with the French sportsman.

Fig Culture and Protection.—Although I have omitted to
speak of some perhaps more important matters, one phase of Parisian
Fig-culture must be alluded to, more as an illustration of patient
and successful guarding against difficulties of climate than anything
else. The trees are trained in long sweeping shoots pretty near the
ground, and in such a form that they may be readily interred in the
ground when the winter and its dangers come. Figs. 62 and 63
represent the aspect of such trees on level ground. I will not enter
into the details of the culture, but simply show how this style of
Fig-tree is protected, and how adapted to sloping ground. The

Fig. 62.

frosts are often of great severity in the neighbourhood of Paris; so
great indeed that the Fig would have little or no chance if left exposed. So in autumn the sagacious cultivators throw the branches
into four bundles, make a little trench for each, and cover as shown
by Fig. 64, with small sloping banks of earth, protecting the crown
of the root by means of a little cone of earth, which merges gradually into the four little ridges that protect the branches. However, all this is shown in the accompanying figure.

When the plantation is made on deeply inclined ground a somewhat different system is followed. The modification is shown by

FIG. 63.

FIG. 64.

FIG. 65.

Figs. 65 and 66.   A basin is formed below the root, so that the water
which would otherwise run off the slope be retained for the greedy

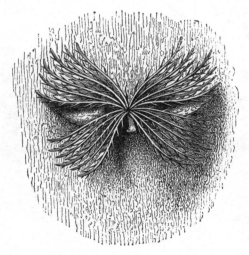

FIG. 66.

roots. As the Fig may be seen in good bearing condition as a standard tree in Sussex, it has occurred to me that this mode of culture might be practised with some advantage on our sunny railway banks in the south, which are now naked and utterly useless.

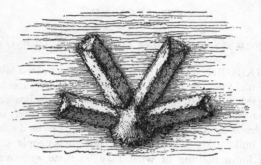

FIG. 67.

In interring the sloping plantations a slightly different system is pursued—the branches are conducted in an upward direction, and covered in, as shown by Fig. 67.

CHAPTER XIV.

A New and Cheap Method of making Garden Walls.

 GARDEN without good walls well covered with fertile fruit trees is in our climate a very imperfect sort of garden indeed, if the production of first-class fruit be a chief object with the owner. It is true that of late years some few have declaimed against garden walls, and said, Let us do away with them and build glass houses instead ; but such people are chiefly those who know nothing of what may be done with walls by a good cultivator, and those who have completely failed from want of skill in making anything of their own. The advice may be good in cold northern districts and places where fruit cannot be ripened against walls, but to recommend doing away with garden walls generally in this country is simply folly. Walls make up for a bad climate when the trees on them are well managed ; but if neglect occurs, and those shoots or branches which should be so spread as to enjoy all the light and warmth possible are allowed to grow forth wild, and perhaps receive but an annual rough-pruning, or whatever else may be the form of the neglect, it is clear that a garden wall soon becomes an eyesore instead of an object of beauty and utility. And I question very much if anything connected with rural life presents such a combination of the two last qualities as a well-managed garden wall—covered with flowers in spring, and fruit equally beautiful in autumn. For some years past garden walls have been comparatively neglected by the gardening public gene-

rally, and very much to its loss. This is perhaps one reason why, with all our advances, a stock of really good winter pears is a thing but rarely seen, even in large, expensive gardens ; and there can be no greater loss to a fruit-garden than that. This want extends even to our markets. In London, for instance, if you want a really good pear in January, you can get it, it is true, at a very high price, and in one spot or so in this enormous city ; in Paris you will probably pay a high price for it too, but you may there obtain it in any part of the town where there is a good restaurant. It may perhaps be said that climate makes the difference, but no such thing. The French, to have in highest perfection such fine pears as St. Germain, Winter Nelis, Crassane, and Beurré d'Aremberg, grow them on walls around Paris. We should do the same with these and others, and could produce an equally good result by doing so.

Let us pass by the important subject of winter pears for the present, and just glance at the peach question. This fruit attains the finest possible condition when well grown against walls in this country. In others they may grow it freely as a standard tree ; in none can they produce better fruit than may be gathered from walls in England and Ireland. France has very diverse climates, some in which the peach grows well as a standard ; but the best peaches she grows are gathered from walls in those parts where the climate is very like our own. There can be no doubt whatever about the fact, that if we pay as much attention to the peach as the cultivators of Montreuil do, we can attain quite as good a result. It cannot be too widely known that no fruit-tree nailed against walls furnishes a more certain and regular crop than the peach when well treated. Yet, what is the fact? Why, that except a small place known as Central-row, Covent Garden, where you pay from 8*d.* to 1*s.* for an eatable peach, and 3*d.* and 4*d.* each for worse than are sold in Paris for one halfpenny each, and fit for nothing but pig-feeding, London may be said to be without this delicious fruit. And there are thousands of private places very badly off for it also. Our good gardeners understand its culture well enough, but of late years public attention has, by various means which need not be detailed

here, been called away from the fact that, with walls, we can produce first-class fruit, and without them do little or nothing with the
choicer kinds. The " power of the climate" about Paris may be
considered very wonderful by some people, but there is one thing
which it cannot do, and that is, produce better peaches than I
have often gathered from walls both in England and Ireland. Over
the greater part of the country, without question, the peach may be
grown to the highest degree of perfection, and yet, though few
Englishmen could manage, as Johnson did, " seven or eight large
peaches of a morning before breakfast began," they may well say
with him that getting " enough" of them was indeed a rarity. It
is stated in a recently published book on fruits that for the majority
of the population to partake plentifully of peaches, " the only hope
that can be held out involves nothing less than an emigration across
the Atlantic!" The present state of matters justifies the writer in
the remark. The quality of the peaches sold at the lowest, but by
no means a low price, is such as to prevent anybody making a
second investment in them, and therefore the fruit is, as the
writer remarks in describing it, " a luxury confined to the wealthy."
Before it is otherwise, good fruit must be sold at a price that will
put it within tasting reach of others than those provided with a
powdered footman to convey it from the fashionable fruiterer's to
the carriage waiting at the end of the " Row."

I assert that we can grow the peach as well and as cheaply as the
French by paying it special attention in the market-gardens, while
it should be planted to a much larger extent in private gardens and
receive increased attention therein. And so of our other better
fruits. But walls are expensive in construction, and therefore
the general adoption of this new kind will be productive of much
good. However, my present object is to point out, not so much
the advantage of walls, as the fact that for the future they may
be made very much cheaper and better than ever by the adoption
of Tall's apparatus for making cheap concrete walls, as one vital
objection to walls has hitherto been their expense. From the
first moment of seeing that very simple and excellent method in

practice, I was convinced of its great merit for the construction of capital garden walls. In many soils the gravelly subsoil may be quickly thrown into the form of walls; on all clays " ballast" will do equally well, and while we are burning it for the purpose of making garden walls, we may at the same time do so for the sake of ameliorating the texture of the soil, and for walk-making, &c. In many places the very clinkers and like rubbish that people do not know what to do with may be used for this purpose, and that by the aid of the roughest labourers, under the superintendence of an intelligent gardener, who will quickly become acquainted with the working of the method. Thus in many places, where the expense may have deterred people from surrounding their fruit and vegetable gardens with walls, they may be erected without such fear; and in the numerous gardens *inclosed* by walls, but not half sufficiently furnished with them, cross walls may be made (as at Frogmore and other large gardens), to throw them into convenient sections. They will thus be sheltered efficiently, while enough of wall space may be had to plant every suitable kind to furnish us with a splendid crop of the finest fruit, and, more than all, to make it worth our while to devote skilled labour to the walls at all seasons when they require attention. This last is a greater advantage than may appear at first sight. To have well-managed wall trees we must have a good and intelligent gardener to attend to them. Now, in numerous gardens, there is not wall space enough to make it worth our while to do this, and the result is that trees get neglected, and consequently nearly useless. The head gardener has so much to do with houses and many other things, that his time cannot be given to wall trees; he cannot afford to spare a man to attend to the walls at all times, and therefore they become profit-less. But where there is a good deal of wall surface, and the pro-prietor or head gardener will make it the duty of one or more in-telligent men, who have had some instruction in the subject, to attend to the trees upon it, it will be found very profitable to do so; and there can be no doubt that by paying full attention to walls our choice fruit supplies might be enormously increased over the greater part of England and Ireland.

I first became acquainted with Tall's method from hearing of its advantages in building cheap houses for the working classes of Paris, and thus wrote of it at the time in the columns of the *Field :*—

"The providing of decent dwellings for the working classes both in town and country is a very important question, which the French are creditably endeavouring to solve, as may be seen by several of their model houses shown in the park of the Exposition. They are very pretty and convenient, though not so well planned nor half so cheap as houses of a nearly like, but better pattern, now building for the Emperor by contract. I succeeded in discovering these the other day in an outlandish boulevard near the Bois de Vincennes, and found them most interesting, particularly as regards the construction of the walls, which are not of brick or of stone, but simply of what is called ' concrete,' and which is really nothing but rough stones, say from the size of a bean to that of an egg, mixed with a little Portland cement and sand. The coarse gravel or stones are found on the ground, mixed in a rough way by the commonest and cheapest kind of labourers, and pitched immediately afterwards in between two boards set to the size of wall required. This rough-and-ready cement or *béton* sets into a firm wall in the course of twenty-four hours, then the framework of boards between which the material has been placed is elevated and readjusted, the ready-made material is thrown in again, and so they knock up those really useful houses in a very short time without the aid of bricklayer, stonemason, or any of the usual complications of building. Each family is provided with two rooms, a little kitchen (better and more roomy than the one in the park), a water-closet on the English principle, according to the Emperor's wish, and a snug little cellar below ground. The doors, windows, and all woodwork have been cheaply made by machinery; gas and water are laid on, and the floors are of the same kind of concrete as the walls, laid thinly between the iron joists, and boarded over, of course. The plan is the Emperor's own, and is very simple, the only apparent deficiency being in the ventilation, for which there is no provision except the chimneys. Workmen are so fond of surrounding themselves with the fragrance of tobacco and other vapours, that the providing of some small aperture for the escape of heated and foul air near the ceiling is very desirable. The buildings are fireproof, in consequence of the layer of concrete on the floors. The facility and ease with which these buildings are erected is quite exceptional. The mode of constructing the walls is that of Mr. Tall, an Englishman, and deserves the attention of all interested in the matter. Not only rough, useless gravel may be utilized in the erection, but common burnt brick-earth or ' ballast' does equally well, and that is made ready for use by having the very fine sifted out of the rest; it is mixed with about one portion of Portland cement to eight of ballast. The great masses of clinkers that are thrown out from the furnaces in the manufacturing districts as useless are about the very best thing that could be employed for this purpose, not to mention other things to be had for nothing, or almost nothing. The rent is not settled yet, and I forget the anticipated amount, but it is something surprisingly low, considering the convenience of the apartments. The windows opening on the staircase, &c., are made with an iron ornamental grating on the outside, so that the glass may be removed in early summer without difficulty, and replaced in the winter when the cold comes; and this affords a good plan of cooling and ventilating the houses in the hot weather. Though so simple in design, the houses are as pretty as could be desired for the purpose, the contrast between the grated windows and the plain offering some variety; and when the little colony is completed—there are about forty houses constructed and constructing here—it will look quite smart. There has been some doubt expressed as to the stability of these walls; but, upon examination, they seem as hard as stone itself. This kind of structure may be put up much cheaper than the cheapest brickwork; when plastered over, it looks as well as any other, and has advantages for resisting wet or damp not possessed by one of bricks. Its merits are worthy the attention of all interested in the matter of providing decent dwellings for the humbler classes—and who are not? The buildings alluded to will be found worth examination, and are situated in the lower end of the Avenue Daumesnil. If useful for this purpose, why not equally so for out-offices, garden, and many other adjuncts of the country house?"

Every part of these houses was run up with the utmost facility upon this system; and of course the building of garden walls upon it is even much more simple and facile—in fact, nothing can be more so. In the *Field* of 17th August, 1867, it is stated that " all kinds of building materials have almost doubled in value within the last five years, while the wages of workmen have in some instances gone up as high as 33 per cent." Such is the condition of things that has brought forth a remedy.

"It is well known that, in all brickwork buildings, the bricklayer's share comes to a very large sum; and more especially is this the case with respect to stabling, cottages, farm buildings, and wall fencing. Anything, therefore, that will diminish this cost, giving equal stability and bearing power, will produce a most material effect in the construction or reconstruction of such buildings as we have mentioned above, particularly in farm buildings, the cheaper class of houses, and labourers' cottages. Taking the present price of bricks at an average of thirty shillings a thousand, and bricklayers at from five shillings to six shillings a day, some idea may be arrived at of the price which a landowner is likely to pay for improving the property on an estate. When the plan which we are about to describe has been fully developed, it must produce a great deal of good in humbler dwellings.

"As a remedy for this state of things, a Mr. Joseph Tall has patented an apparatus for the purpose of building strong substantial walls, without either bricks, building stone, the bricklayer, or the mason! In fact, it closely resembles concrete, which has for a long time been known in connexion with foundations; but, on account of the very liquid state in which concrete must be used, and its not standing wet, wall-building with it has been hitherto out of the

question in any first-class way. When used for foundation purposes, it is, as we all know, mixed on the spot and tumbled into the
foundation-cuttings immediately, where it hardens in a very short
time. In the present process the composition is Portland cement,
gravel, and brick-rubbish, or broken stones, sea shingle, quarry
refuse, or any other hard, durable substance. In preparing to
build, the foundations are laid with concrete in the usual way,
and, when that has been done, the framing for the wall is commenced. This may be taken to be neither more nor less than a
wall-mould, which is so fastened as to screw together and unscrew
when required."

And if so important for out-offices, &c., how much more so for
garden walls! The great expense of brickwork at the present day
must effectually prevent most people from getting the wall surface
they require to produce abundance of fruit, and therefore this
system is worthy of our best attention. Mr. Tall's patent apparatus
is certainly by far the best means of moulding the concrete; and as
no scaffolding would be required for the walls, their construction
would be a matter of slight expense in most places. For further
information I cannot do better than refer the reader who takes an
interest in the matter to the pamphlet published on the subject by
Mr. Tall, Falstaff-yard, Kent-street, Southwark; adding, however,
the following letter from Mr. W. E. Newton, C.E., who has tried
the system for garden walls. He says—

"The first garden wall I built of concrete was about nine feet
high from the footings. It was nine inches thick, with a coping
also in concrete, and stuccoed on both sides. It has stood well
and has no buttresses. I have also built a wall of seven inches
thick, about the same height, and a boundary wall of only six inches
thick. If I were to build walls for any one else, I should recommend only six-inch walls for any walls not exceeding nine feet in
height. I am not sure we might not venture to twelve feet, but
as that would involve scaffolding which is fixed to the wall, it
would perhaps be better to adopt nine-inch work, at any rate for

the lower part. A coping might be made of slate slabs (Fig. 68), or it might be made of concrete. I should prefer it in concrete with

FIG. 68.   FIG. 69.

flat iron bars running across to strengthen it, as shown in Fig. 69. I would undertake to build a wall of this kind for four shillings per yard super, if materials are convenient to be had."

It may be stated, in order to prevent misapprehension, that Mr. Tall makes no claim of novelty in using concrete for walls. There are in existence ancient churches made of this material, as well as buildings of recent date; but the difficulty of construction, with boards for moulding the walls, has hitherto been the great bar to its use. The novelty introduced by Mr. Tall is the moulding apparatus, which at the same time serves for the scaffold, and greatly cheapens the process of erection; and it is for this that the patent is taken out. It will be better to insert the irons and eyes for the support of the galvanised wires while the successive layers of the wall are being built up or while the material is soft.

CHAPTER XV.

Culture of the Orange (by M. Hippolyte Jamain)—Oleander Culture (by M. Chaté fils).

[M. Jamain of Paris is probably the largest and most successful grower of the Orange Tree in any northern climate; and I have much pleasure in offering the following article by his son, convinced that the system described is the sound one for England.—W. R.]

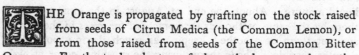HE Orange is propagated by grafting on the stock raised from seeds of Citrus Medica (the Common Lemon), or from those raised from seeds of the Common Bitter Orange. For the trade, plants grafted on the lemon stock are the most suitable, the lemon growing more vigorously than the wild orange tree; but to secure the plant long life, the latter is the most preferable. The reason of this will be easily understood; the difference between the lemon and orange trees is much the same as between the quince and wild pear trees : like the quince, the lemon makes all its roots at the surface of the soil, the wild orange goes deeper, and consequently the tree is better able to resist the wind and the vicissitudes of the season ; naturally there is more analogy between the two woods, and the result of experiments is that the plants live much longer. An orange tree grafted on the lemon may live about a hundred years ; after that time it decays and perishes ; an orange grafted on its wild congener may live over 300 years— witness the Grand Bourbon in the Orangery at Versailles, near Paris, which tree is now more than 400 *years old, and is grafted on*

the Wild Orange. Sow the seeds early in the spring in a light but not too sandy soil, and in pots (twenty-five to thirty per pot) ; put the pots upon a dung-bed (lukewarm), and keep the soil fresh, but do not have any steam in the frame, and to prevent this give a little air (one-half inch) to the front lights. When the seeds have come up, encourage them to grow to three or four inches high. Afterwards put them in a warmer bed, and keep a damp warm atmosphere in the frame ; shade them against the burning rays of the sun ; and when they are seven or eight inches high, give them a little air, increasing it as they get stronger. Let them pass through the winter in a greenhouse, where the temperature must not descend lower than 40° Fahrenheit, and in early summer put them on another hotbed in the open air plunged in leaf mould or cocoa fibre. Leave them plunged on this hotbed through the summer, and give them plenty of water, and from time to time a little liquid manure. About the end of August in the same year graft them by the same method as that practised for roses in the winter, and put them on a hotbed, keeping as much damp vapour about them as possible. Shade them during the sunshine, cover at night and keep them close as long as the grafts are not well united together ; they will be safe long before the early frost. Keep them in the frame during the winter, and the next spring divide and pot them in rich light soil mixed with a very little silver sand to prevent the soil becoming hard : put the pots on a hotbed in a frame, and after they are rooted give them plenty of air. In the middle of June, make a hotbed in the garden and put them on it without any covering whatever, giving plenty of water during the hot weather, and give them three or four times a little liquid manure to encourage active growth. Before the first frost they must be housed, and they will do through the winter in a green-house where the temperature is kept three or four degrees over the freezing point.

During the spring of the following year pot the plants afresh, and place them on a hotbed covered with a frame ; keep them closed until the roots begin to shoot, and give air successively ;

shade the frame against the burning rays of the sun, and when frosts are no longer to be feared, take the lights off entirely. When they have done their growth, and the wood is sufficiently ripened, pot them afresh, and leave them in a greenhouse for a week or two. In June make a hotbed in the open air, covered five or six inches with dung-mould or cocoa refuse, and put them in it. This is the last season during which the oranges need be grown upon hot dung bed. The greatest obstacle to the well growing of the oranges in England is the persistence of the gardeners and nurserymen in treating it as a greenhouse plant. I do not mean to say the oranges should be left like our common shrubs, but it is possible, with very little care, to grow them in England almost as well as in northern France. Many writers on this subject give the south exposure as the best for an orangery, and therein is the mistake. To insure the success of oranges grown in boxes or in pots, *they must not in any cases be allowed to grow in the houses;* all their growth must be made *out of doors;* and it is a matter of fact, that if the orangery is to the south, no matter what the trouble you take to prevent their starting, the plants will be beginning to shoot a long time before the weather is mild enough to permit of their being placed in the garden. A good orangery should have a northern exposure, with plenty of windows to admit the light, and every convenience to give full air when it is not frosty. It will be very easy to heat the orangery in such a position, as the temperature required is only two or three degrees over the freezing point. It must be remembered that oranges are grown out of doors all the year round in parts of France and Spain where it freezes every winter. If the plants, after all the care taken to prevent their growth in the houses, begin to vegetate, and if the young shoots are more than an inch in length, it would be far preferable to cut them back than to let them retain a growth which is sure to be disfigured and spoiled in the open air. The watering of the oranges must be very carefully done, as too much water would be more pernicious than too little, and especially for the large plants, where the soil is in greater quantity; one or two injudicious waterings are enough to kill the best established plants. Good drainage in the bottom of the box or pot will prevent many accidents. In the winter they want very little water. Before watering them, the grower should feel the leaves of the tree, and if flabby, as though on the point of flagging, it is then time to give them water. This applies only for the large plants, the large quantity of soil employed for them keeping its moisture for a long time. The small plants must be watered more frequently, but still with great moderation in winter. During the summer water must be given freely, but not in excess. The best time to do the watering is in the morning; and at night the plants will require a little syringing on the leaves, but only in the hottest time of the year. Liquid manure given with great moderation will do them good and quicken their vegetation. The small plants which have passed beyond the hotbed stage should be potted in a very rich light soil, and not too sandy, say nine parts of soil divided as follows:— Three of maiden loam, two and a half of yellow loam, one and a half of old dung mould, one of peat, and one of sand. In potting plants of a larger size, the soil should be a little stronger, and be composed as follows:—Three and a half of maiden loam, three of yellow loam, one of thoroughly rotten dung, a quarter of peat, and one part of sand.

OLEANDER CULTURE.—Visitors to the Continent in the summer months can hardly fail to be struck with the employment of certain plants for decorative purposes, of which we in this country make comparatively little use. Here, if a few orange trees or Portugal laurels, perchance a pomegranate, are grown in tubs and put on to the terrace in summer time, it seems to be considered that enough has been done in that way. There is no reason, however, why many other plants should not be used in like manner. We well remember the beautiful effect produced on a quay fronting the lake of Lucerne by a number of standards of this kind, including not only the plants mentioned, but Pittosporums, Yellow Jasmines, Evergreen Oaks, Euonymus, Aucubas, and Figs. At

Vienna a similar assortment may be seen in front of some of the principal cafés, where one may sit in the open street under the shadow of the pomegranate and the oleander. This latter plant, too, is an immense favourite with the Parisians. In fact, the oleander forms, with the myrtle and the pomegranate, one of the most important articles of Parisian commercial horticulture. The reasons for this are obvious—the elegant habit, glossy foliage, profusion of bright rosy or white flowers, endowed, more-over, with an agreeable almond-like perfume, offer recommendations hardly to be exceeded by those of other plants. The culture, moreover, is easy. Indifferent as to the treatment it receives in winter, it may be kept in cellars or garrets—almost anywhere, in fact; hence its frequency abroad in the windows of the artisan and at the doors of the merchant's office The shrub may be propagated either by layers or by cuttings; but of late years, in France, the former method has been abandoned, as it is found that cuttings produce plants of better habit, and in greater numbers. It was from cuttings that those beautiful little oleanders exhibited in the reserve garden of the late Paris Exhibition, in the first fortnight of June last, by M. Chevet, were obtained.

A well-known French horticulturist, M. Chaté fils, who has had great experience in the cultivation of this plant for the last twenty-five years, has obligingly communicated to us the details of his method of cultivating this plant, which are as follows :—" If layers be required, about the end of April or beginning of May, a period when in Paris greenhouse plants are placed in the open air, some old stocks of oleander are planted out in the soil, previously trenched and well manured. At the end of a month these stocks are rooted in the soil, and then all the branches are bent down to the ground and fixed in that position with pegs. They are then covered with soil about fifteen to twenty centimetres (six to eight inches) deep, taking care to leave the extremities uncovered. The branches are copiously watered, especially when the roots begin to be formed. In the early part of August the branches are cut through, and thus separated from the parent stool without disturbing the roots. About the beginning of September the young plants are placed in pots proportioned to their size. Formerly gardeners used to layer at once into pots, thinking that by adopting the method thus described they would lose their plants. They also used to cut the branches, as they would do with pinks, at the places where they wished to secure the formation of roots. But after a time, it was found that this practice was unnecessary. It was even seen to prevent the free development of the plant, which thus became less vigorous than when layered in the soil. Although layering secures the formation of strong plants quicker than propagation by cuttings, the plants are not of such good habit, and are not multiplied so rapidly as by cuttings. These may be obtained at any season when the two principal requisites for the formation of roots—heat and moisture—can be secured ; the best cuttings, nevertheless, are those raised in April, in pots or pans filled with good light soil mixed with a little peat. The cuttings should be taken with little shoots obtained from the two-year-old wood, and the whole should be placed on a brisk hotbed. Under such circumstances the cuttings will root in from fifty to sixty days. As soon as they are rooted they are exposed gradually to the air; they are repotted so that they may be placed during the summer in the open air. As soon as the plants attain a height of about twenty inches the shoots are stopped in order to induce the formation of flowering branches. To form a clean little stem, all the buds formed on the lower eight inches of its base are suppressed. The shrub especially delights in copious waterings and should be kept perfectly free from the insect pests which infest it when under negligent cultivation."

The above is part of a leading article from the *Gardener's Chronicle.* In addition I may remark that pretty little free-blooming oleanders are grown about Paris in small pots, say 48's, in sandy soil, and these pots they of course soon fill with roots. They are plunged all the summer in the open air, and grown at all other seasons near the glass in those low houses so much in vogue in Parisian nurseries and gardens. The large plants you see in some of the public gardens are in great tubs, evidently undisturbed for long periods. They flower profusely, and get about the same treatment as orange trees, as regards housing in winter. The little plants of oleander are, however, most likely to be useful with us. They are allowed to rise with an undivided stem for about four inches, and then break off into several branches. There should be no difficulty in growing them wherever there is a sunny shelf in the greenhouse, by securing a clean, while discouraging a soft or luxuriant, growth, giving a rather dryish rest in winter and abundant water and light in summer. In winter any cool house will do to store them, or even a shed. The large round-headed plants in the public gardens are certainly very noble objects, and more worthy of culture than the orange tree tubs. Judging by the habit of the oleander, as generally seen with us, it might be supposed that they would not make ornamental trees for a terrace, but nothing can be finer than the immense specimens seen in the Luxembourg Gardens, the heads being as round and dense as a Pelargonium grown by Mr. Turner, and sometimes so much as ten feet through; and as for the little plants grown in six-inch pots, nothing can be prettier.

CHAPTER XVI.

Asparagus Culture in France.

ASPARAGUS is grown much more abundantly and to a much larger size in France than it is in England. The country is half covered with it in some places near Paris; small and large farmers grow it abundantly, cottagers grow it—everybody grows it, and everybody eats it. Near Paris it is chiefly grown in the valley of Montmorency and at Argenteuil, and it is cultivated extensively for market in many other places. About Argenteuil 3000 persons are employed in the culture of asparagus —at least so my informant told me, and he is the son of the cultivator who has taken the best prizes for asparagus at the Exhibition. His father not being at home, I traversed a considerable portion of what may be termed the region of asparagus with this youth, who was of the intelligent type, and understood all about this dainty vegetable. We first saw it growing to a large extent among the vines. The vine under field culture, I need scarcely say, is simply cut down to near the old stool every year, and allowed to make a few growths, which are tied erect to a stick: they do not overtop the asparagus in any way, but on the other hand the strong plants of that showed well above the vines. It was not in distinct close lines among the vines, but widely and irregularly separated, say six or seven feet apart in the rows, and as much or more the other way. They simply put one plant in each open spot, and give it every chance of forming a capital specimen, and this it generally does. The great "stools" appear to be irregularly scattered about here and there. When they get large and a little top-heavy in early summer, a string is put round all, so as to hold them slightly together (the careful cultivator uses a stake), and the mutual support thus given prevents the "grass" from being cut off in its prime. We all know how apt it is to be twisted off at the collar by strong winds, especially in wet weather, when the drops on every tiny leaf make the foliage heavy. The growing of asparagus among the vines is a very usual mode, and a vast space is thus covered with it about here. But it is grown in other and more special ways, though not one like our way of growing it, which is decidedly much inferior to the French metnod. Perhaps the simplest and most worthy of adoption is to grow it in shallow trenches. I have seen extensive plantings that looked much as a celery ground does soon after being planted, the young asparagus plants being in a shallow trench, and the little ridge of soil being thrown up between each. These trenches are generally about four feet apart. The soil is rather a stiff sandy loam with calcareous matter in some parts, but I do not think the soil has all to do with the peculiar excellence of the vegetable, and am certain that soils on which it would flourish equally well are far from uncommon in England. It is the careful attention to the wants of the plant that is so beneficial. Here, for instance, is a young plantation planted in March, and from the little ridges of soil between each shallow trench they have just dug a crop of small early potatoes. Now, in England, those plants would be left to the free action of the breeze, but the French cultivators—like the old Scotchwoman who would not trust the stormy water and God's goodness as long as there was a bridge in Stirling—never leave a young plant of asparagus to the wind's mercy as long as they can get hold of a bit of oak about a yard long. But when staking these young plants they do not insert the support close at the bottom, as we are too apt to do in other instances, but at a little distance off, so as to avoid the possibility of injuring a fibre; and thus each stake leans over its plant at an angle of 45°, and when the sapling is big enough

Q

to touch it or be caught by the wind, they tie it to the stick as a matter of course. The ground in which this system is pursued being entirely devoted to asparagus, the stools are placed very much closer together than they are when grown among the vines, say at a distance of about a yard apart. The little trenches are about a foot wide and eight inches below the level of the ground—looking deeper, however, from the soil being piled up.

They place the young plants in these trenches very nicely and carefully. A little mound is made with the hand in each spot where a plant is to be placed, so as to elevate the crown a little and permit of the spreading out of the roots in a perfectly safe manner. In fact they seem to be about as particular as regards depositing the young plants in the first instance, as a good grape-grower is about his young vines. They plant in March and April—using any kind of manure that can be had, but chiefly here, so far as I could see, the refuse of the town—the ashes, old vegetables, rags, and other matters, that the people throw before their doors, and which the dust-carts take away in the morning. They are very particular to destroy the weeds, and they also take good care to destroy all sorts of insect enemies in the early morning, especially during the early summer. Between the lines of asparagus they plant small growing crops on the little ridges during the first years of the plantation, but are careful not to put the large vegetables there, which would shade and otherwise injure the plant. When they plant they spread a handful or so of well-rotted manure over each root, and, so far as I could make out, they repeat this every year, removing the soil very carefully in the autumn down to the roots, putting on them a couple of handfuls of rotten manure, and spreading the earth over again, so that the rain is continually washing manure to the roots. When doing this they remark the state of the young roots, and any spot in which one has perished, or has done little good, they mark with a stick, to replace it the following March. Early every spring they pile up a little heap of fine earth over each crown. When the plantation arrives at its third year they increase the size of the little mound, or, in other words, a heap of the finest and thoroughly ameliorated earth is placed over the stool, from which some, but not much, asparagus is cut the same year, taking care to leave the weak plants and those which have replaced others "abandoned to themselves" for another year. They would appear to cut the best of it when it is about an inch and a half out of the ground—and here is the only objectionable thing about their system. The top is very good, but as a rule too short ; but such a handle as they give you to it ! Now, it is very desirable to have something to take hold of, but to cut it as they do here, and as we often do in England, is not wise, nor conducive to the thorough enjoyment of the vegetable ; but then it is simply a matter of the amount of covering given, or of the depth at which it is cut, and therefore of the simplest management. The care and culture may be applied as described, and the asparagus cut at pleasure. To procure it in a thoroughly blanched condition, the French pile up these little mounds of fine earth, which enables them of course to get it much longer ; besides, they can pull away the soil conveniently, and get at the rising stems as low down as they like. It is not, however, the fault of the cultivator that the asparagus is so much blanched, for I have been told by the first fruit and vegetable merchant in Paris that his customers would not buy the finest Asparagus ever grown if brought in in a green state. This is why you see it with a shaft like ivory and with the point of the shoot of a red, rose, or violet tinge. When the plantation reaches its fourth year the little mound of blanching earth is increased to fifteen inches in height, for then they expect to cut something worth while, and these mounds are made in the early part of March ; and even after this, as they grow stronger, the little mounds are increased ; and they always keep a look-out for the feeble plants, with a view to replace them.

To have asparagus as it ought to be, they say, you must cut every day, or every two days, according to temperature, so that it may be obtained at the right moment ; indeed, if they do not do this, the shoots become too high and too green. They place great importance on obtaining strong and healthy plants ; and in the establishment which I visited they have three kinds, l'Ordinaire,

La Hollande tardive, improved, and La Hâtive d'Argenteuil. The first is described as very fine, the second very strong, and the last as the earliest, most productive, and best. Of course there are various modifications of the plan described herein, and in several instances I saw two rows placed in a rather wide trench in an alternate manner. As to the size and quality of the asparagus produced by this method, there can be but one opinion. Mr. Veitch and many other English horticulturists, who know what gardening is, as well as it is possible to know it, have been, with myself, surprised at it. The same difference holds good in the forced asparagus—the slender pipe-shank productions of the English forcing-house being miserable compared to it.

To sum up: the French mode of cultivating this delicious vegetable differs from our own diametrically in giving each plant abundant room to develope into a magnificent stool, in paying thoughtful attention to the plants at all times, and in planting in a hollow instead of a raised bed, so that as the roots grow up they may have annual dressings of enriching manure. Every year they lay bare the roots, carefully spread good rotten manure over them, and then add an inch or two of soil over that. And in this way they beat us with asparagus as thoroughly as Messrs. Meredith, Henderson, or Miller, beat them with hothouse grapes. A man who knows how to spend two and a half francs for his dinner in Paris enjoys asparagus much longer and of much better quality than many a nobleman in England with a bevy of gardeners. In the first-class restaurants you usually pay high for asparagus, as you do for all other vegetables, but it is served very cheaply in many respectable ones—so much so, indeed, that it is partaken of by all classes.

As the culture of this vegetable is so important, and the French manage it so well, I venture to go further into detail by giving the following, written by a well-known and very successful cultivator of Argenteuil, and which first appeared in the *Gardener's Chronicle*. I have made some little alterations, with a view to rendering the meaning simpler and clearer to the reader :—

"*Preparation of the Ground.*—When a convenient piece of ground has been selected, it is first of all to be mellowed by spreading on its surface a good dressing of horse or sheep dung. The lowest layer of a dunghill, the dregs of grapes, and night-soil, are likewise good manures. The ground is to be dug up to a depth of sixteen inches in fine weather at the beginning of winter, during which season it is to be left at rest.

"In the month of February following, at least as soon as severe frost is no longer to be expected, the ground is to be laid out in furrows and ridges, in order to shape the shelving beds, and the excavations which are to receive the plantations. For this purpose the following operations are to be performed:

"First, there are to be drawn the whole length of the ground, and by preference from north to south, two lines, leaving between them a space of 14 inches, intended for the site of the first half-shelving bed. Reckoning from the interior base of this half-shelving bed, a distance of 24 inches is to be measured for the first 'ground' or trench. The earth taken from it will serve to form the shelving bed. The second shelving bed, which will be a complete one, is to measure 28 inches in width at its base, and 14 inches in height. Next comes the

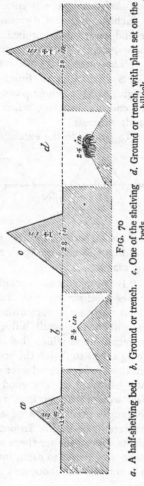

FIG. 70

a. A half-shelving bed. *b.* Ground or trench. *c.* One of the shelving beds. *d.* Ground or trench, with plant set on the hillock.

second trench, then the third entire shelving bed, and so on, until the whole piece of ground has been occupied. Thus, the first half-shelving bed will measure in width 14 inches, and in height 8 inches; the first 'ground' or trench in width 24 inches, the second entire shelving bed in width 28 inches, and in height 14 inches, &c. (See the preceding diagram.)

"The earth of the shelving beds being intended to cover over the plants from time to time, these beds will gradually diminish in height, and the whole piece of ground will become nearly level at the end of five years, when the asparagus plantation will be in full productiveness."

[In justice to the extensive market grower and successful prize-

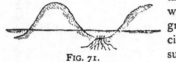

FIG. 71.

taker who thus describes his culture, we are bound to respect his diagram; but a readier and less precise method is more generally pursued, such as that indicated by Fig. 71, roughly drawn from memory.]

"*First Year.*—The first plantation is to take place during the months of March or April, and should be performed in the following manner:

"In each trench, through its entire length, small holes, eight inches in diameter and about four inches deep, must be formed about thirty-six inches distant from each other. In the centre of each of them a small hillock of earth about two inches high is to be raised, upon which the asparagus plant is to be laid down, care being taken to divide the roots equally in every direction; the roots are then to be covered over with half an inch of earth; and one or two handfuls of very good manure are to be added, and covered over with about an inch and a half of earth, at the same time forming a small hollow of about an inch deep over each plant, to indicate its position. In order properly to know the position of the plants, and to shelter them and their shoots from accidents, a small stake is to be set to each, inclining it at an angle of 45°, in order not to injure the roots, and placing it a little away from the plant.

"Every year towards the months of April and May slugs and snails are to be carefully looked for, while the morning dew remains, and destroyed. Beetles are also much to be feared in the asparagus plantations. Twice every day during a fortnight it will be well to pursue these insects with rods, so as to hinder them from depositing their eggs on the stalks of the asparagus; these eggs appear at the end of three weeks in the shape of black maggots or worms, which prey upon the asparagus stems and dry them up. Yet these insects are not the only ones which are to be dreaded. The white worms (or maggots of tree beetles) are very dangerous, and it will be well constantly to put in use the most proper means to get rid of them, for they eat the roots and destroy the asparagus plants. It will be useful also to set mole traps, for while tracing their underground roads the moles cut the asparagus roots in order to get through. Frequently during the season the plantations should be thoroughly cleaned, taking care to never bruise or in any way injure the young plants, for any accident to these is of course directly prejudicial to the plant.

"Common vegetables, such as late potatoes, cabbages, &c., ought not to be planted on the ridges of beds, which, however, may be made useful (but only during the first years) by growing on them early potatoes, lentils, kidney beans, salads, and such other vegetables as are of little inconvenience from their dimensions.

"In the month of October, during fine and dry weather, the small stalks of the asparagus are to be cut off at six inches above the ground. The ground is to be lightly cleaned, and the shelving beds must be dug up to a depth of twelve inches, maintaining their conical shape.

"The asparagus is to be lightly covered with manure, the plants being laid bare with a flat hoe, for a diameter of eight inches, and up to the crowns. Proper care ought to be taken not to injure the roots with the implement. On each plant lay one or two handfuls of good manure, free from all noxious substances. While spreading the manure, mark out with a small stick the site of the plants

which have failed during the course of the year ; these must be re-
newed in the month of March following.

"The manure is at once to be covered over with about three
inches of the best mellow earth at hand, and over the plants is to
be made a small conical hillock about two inches high. This
operation is the last to be performed for the year.

" *Second Year.*—In March or April begin by replacing the plants
which have failed in the preceding year, selecting vigorous plants a
year old, and setting them in the same manner as recommended for
the first year. Stakes are to be placed near the foot of each plant,
always at an angle of 45°.

" In the beginning of April a cleaning is to be made on the
shelving beds and on the grounds ; it will be well to perform this
operation the day after a sprinkling of rain, in order the more easily
to break the clods.

" As soon as the asparagus stems become firm, fasten them to the
stakes, in order to protect them against the wind, which might
break them: In the month of October the dry stalks are to be cut
off at eight inches above the ground; the shelving beds are to be
turned up, always lightly hollowing out the trenches. Manure is
to be spread on the shelving beds, which are then to be dug up.
The stakes, having become useless, are to be taken away. Lastly,
the laying bare of the plants is to be done by taking away the earth,
as already directed, up to the surface of the manure. The earth
must be mellowed by the hands, and covered at once over the
plants, to the thickness of a couple of inches, always in the shape
of a small hillock.

" *Third Year.*—In the middle of the month of March, during fine
weather, small knolls, from six to eight inches high, are to be made
over each plant, taking nevertheless as a basis the comparative
strength of the crowns, more or less large, or of a more or less
determined development; those which may be too feeble, or having
served the preceding year to supply the bad ones, or those which
had failed, are to be covered over with a hillock of only four inches
high, and should then be left to themselves."

[It should be borne in mind that the use of these knolls is to blanch the asparagus; but as this is well known to be no improvement to the vegetable, the piling up of little knolls need not be performed except where fashion requires the cultivator to furnish his asparagus in a blanched state. One secondary advantage of the knolls should be stated, however. When cutting the asparagus, the fine mould of the little mound may be easily removed, and the shoots taken off low down, and without any injury to the plant.

" From the other plants, three, or at most four, asparagus heads may be gathered; but they are not to be cut off with an asparagus-knife, but removed with the fingers. However, there is a particular sort of knife, square-shaped at the end, and having teeth on one side, forming a saw, which will be useful to take away the earth about the stalk, and will make it easy for the fingers to reach the subterranean stock, which care must be taken not to injure.

"With regard to the gathering, one finger must be got behind the asparagus stem at its base, and by bending it, it will easily come off the stock. In this manner all injury to its neighbours, which may easily happen with an asparagus-knife, will be avoided; and there will not be left any wounded ends, from which the sap will flow and spread around, occasioning rapid decay. Care should be taken to close up the hole made for the gathering of the asparagus, and the knoll is at once to be formed anew.

"In the month of April, the stakes are to be again used, and the stems fastened to them in due time. After having, in fine weather, done all that is necessary in the way of cleaning, in the month of October the dry stalks are to be cut off about ten inches above the ground, and the dead rubbish thrown out of the asparagus plantation. From the whole surface of the trenches, and to a depth of four inches, the earth is to be taken away and thrown upon the ridges; this earth is to be substituted by a layer of very good manure, which layer is to be of a thickness of about an inch and a half, if night-soil is made use of, or of about two inches if it is only common manure. At the same time a portion of the end of the old stalks is to be taken away, preserving that nearest to the crown, so as to indicate the exact site of the plants for the fourth year.

"After having spread the manure, the ridges must be dug up, and the manure covered with an inch or two of earth from the ridges, a small hillock being left over the crown of each plant.

"*Fourth Year.*—About the middle of March, in dry weather, or the day after a sprinkling of rain, knolls of the height of from ten to twelve inches must be formed over each plant with good mellow earth. The feeble plants, marked with a small stick at the preceding laying bare, are to be covered over with hillocks of a thickness of from four to six inches only.

"While earthing up the asparagus the ends of the dry stalks are to be taken away. The gathering is to take place from the largest ones during one month at the most. Then they are to be left to run into seed. The most feeble ones are to be spared in order to strengthen them. At the second dressing in the month of May, earth is taken from the shelving beds, in order to cover over, to an extent of an inch or two, the whole surface of the grounds, so as to protect the asparagus plantation from the dryness of the summer. The stakes should be five feet high.

"In the month of October the stalks of the asparagus are to be cut off at fourteen inches above the ground, and the plantation is to be cleared of the rubbish; manure is to be spread on the ridges, which are to be made up from the knolls in the trenches; and are then to be dug up to a depth of sixteen inches.

"Notwithstanding the manure laid upon the shelving beds, the roots of the asparagus are to be carefully laid bare in the manner already described. Upon the crowns are to be put a few handfuls of good manure, which is to be covered over with two inches of good mellow earth; the knolls, which are to be formed over the centre of the plants, are to be over three inches in height.

"*Fifth Year.*—The making of knolls on the asparagus is to begin in the month of March; the knolls are to be fourteen inches high, and their diameter is to be in conformity with the dimensions of the plants. The gathering is to consist of the heads on all the large plants, and of some only on the feeble ones; the gathering may last two months at most. In order to get fine asparagus, the heads are to be gathered once every day, or every other day, or every third day at farthest, according to the degree of temperature. This is the way to obtain rosy, red, or violet asparagus. In order to get it green it will be sufficient to let the heads grow during four or five days more; they will lengthen and become green. The second dressing is to be made as in the preceding years. The stakes are to be put in as soon as the necessity is felt, and the stems, having regard to the increase of their height and weight, must be firmly tied, so that the wind may not disturb them and that they may not be broken.

" In the month of October following, the dry stalks are to be cut off at fourteen inches above the ground. The plantation is to be cleared, and the ridges are to be replenished by adding to them the earth of the knolls which have been raised on the plants for the gathering. Then the manure is to be spread in the manner already indicated; and the digging up of the ridges is next to take place.

" *Sixth Year.*—When the asparagus plantation shall have reached its sixth year, it will then be in full productiveness. The forming of knolls is to take place in March during fine and dry weather; the knolls must always be fourteen inches high, reckoning from the subterranean stock.

" Nevertheless, the care to be taken is to be the same as in the preceding years, particularly with regard to cleanliness and staking. As for insects, they will be less to be feared than during the first years of the establishment of the asparagus plantation. The beetles can no longer lay their eggs on the stalks of the asparagus, since they are cut during two months, and when allowed to start up the time of the laying of eggs is past.

" In the month of October the shelving beds are to be turned up in conformity with the manner shown for the preceding year; the shelving beds and the plants are to be manured, as has been explained for the fourth year. As the asparagus plantation may last fifteen or twenty years, the operations and the care to be taken are to be repeated from year to year in the manner above indicated.

" Generally, in a well-established asparagus plantation, the gathering, reckoning from its beginning, is to take place during two months, whatever may be the climatic circumstances under which the plantation is placed.

" It must have been seen that the expense is not very great ; the chief object is the care which must be taken.   The main point is to get good plants, in order to obtain good produce.   By properly following the rules laid down here, satisfactory results will be obtained."

[The mode of forcing asparagus chiefly consists in digging deep trenches between beds planted for the purpose, covering the beds with the soil and with frames, filling in the trenches between each bed with stable manure, and protecting the frames with straw mats and litter to keep in the heat.]

CHAPTER XVII.

Salad Culture.—Mushroom Growing, &c.

NOT only do the French gardeners supply their own markets with delicious salads all through the winter and early spring months, but also, to a considerable extent, those of some other countries, and send vast quantities to the English markets. Now it will probably occur to the reader that climate is the cause of the superiority of the French in this respect, and, indeed, some practical men repeatedly say so. Nothing can be more fallacious than this belief, and I have no hesitation in affirming that, by the adoption of the method to be presently described, as good salads as ever went to the Paris markets may be grown in England and Ireland during the coldest months of winter and spring. It is simply nonsense to say the climate does it; the winters in northern France are severer than our own, and I know many spots in England and Ireland which are preferable to the neighbourhood of Paris for this culture. Near that city I have often seen beautiful cos and cabbage lettuces looking as fresh under their coverings in the middle of winter, when the earth was frost-bound, as the budding lilac in May : had they been treated as ours usually are, they would have presented a very different appearance. At all times of the year the gardens in which salads are grown round Paris are beautiful examples of cultivation. In the spring and summer, when they are grown in the open air, nothing can be more agreeable to look upon; but it is their condition in the cold season, when little or nothing can be done with them out of doors, that demands most attention from us. As very ordinary cultivation suffices to grow them with us in the favourable parts of the year, and in the other our markets are supplied from France, it is obvious that it is as regards the winter and early spring supplies that we want improvement. That improvement is easily secured.

The first and the chief thing to do towards it is to procure some of the large bell glasses (*cloches*) which the French use for this purpose. They are simply huge bell glasses made of cheap glass, with a round knob at the top to permit of their being handled with facility. Not only for salad growing, but for many other purposes, these are among the most useful things that can be employed in the garden. The French shift them from one salad to another, raise seedlings under them, strike cuttings under them, forward asparagus and many other things with them, and, in a word, make more use of them than any other article they use in their gardens. They are cheap—costing about eightpence each if bought in large quantities, and ninepence or tenpence if bought in small numbers, or singly—require no repairs, are easily cleaned and stored when not required. The nice task of giving air is done away with in their case. Without air on "every possible occasion" the British gardener attempts nothing under glass. By adopting this simple article, he may forego that ceaseless trouble throughout the winter and early spring. In the hotter weeks of autumn, these glasses are tilted up on one side for an inch or so, with a bit of stone placed underneath; but when once winter comes in earnest, then down they go quite close, and are all through the winter in the same condition as what we call Wardian cases. By the way, the French recognised this principle long before we did, and what is more, have made a far more practical use of it. For all sorts of winter salad-growing this huge bell glass of theirs is infinitely superior to anything that we use for like purposes. The plants get full light at all times, and, while perfectly preserved from the filth and splashing of the rains in winter, are not in the least "drawn" or injured by the confinement, the light coming in so freely at all points.

The glasses are nearly sixteen inches in diameter, and about as much in height. For the winter work they are placed on a sloping spot with a sunny aspect, or the ground is thrown into beds wide enough to accommodate three lines of glasses. In early autumn these beds are made and the plants placed upon them, so that they can be readily covered by the bell glasses when the time comes that growth is checked in the open air. It should be added that the ground chosen is thoroughly rich, light, and well and deeply stirred, and the lettuces are sown at intervals of a fortnight or so, so as to secure a succession, and to provide for the wants of the various kinds. The plants put out in September for the early and mid-autumn supply may not require to be covered if the weather be fine; and if they do, the glasses are a little tilted up as before de-

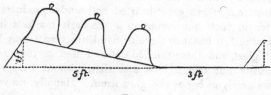

5 ft.   3 ft.

FIG. 72.

scribed. But when the sun begins to fail and the cold rains to check growth, about the end of October, then the crop to be cut in the following month must be covered; and when towards Christmas the frost begins to take hold of the ground, the glasses must be firmly pressed down, and a deposit of leaves and litter placed among and around them. Thus, while all around is at rest in the grip of ice, the plants will be kept perfectly free from frost, while receiving abundant light from above, and growing as fresh as April leaves. Of course a deeper layer of this surrounding litter will be necessary in case of severe frost than is so in early winter. Covering them a little more than half-way up with a rather compact body of leaves and litter, effectually secures them from sharp frosts. When very severe frosts occur, mats made of straw are spread over the tops of the glasses; and should heavy falls of snow occur while these mats are on, they will enable the cultivator to carry it bodily away from the bed or beds; for it should not be allowed to melt on the beds or in the alleys between. In late spring the *cloche* is not required, nor is it for any except those crops that require bringing forward. Thus the March and April supply is planted in October on a nice light bed, with a surfacing of an inch or so of thoroughly rotten manure or leaf mould. These little plants are allowed to remain all through the winter unprotected; and when in spring the most forward lettuces are cut, the glasses are immediately placed over the most advanced and promising of the little ones that have remained exposed. By that time they have begun to start up, encouraged by the early spring sun, and from the moment they receive the additional warmth and steady temperature of the *cloche* they commence to unfold the freshest and most juicy of leaves, and finish by becoming those great-hearted and tender products which one may see in such fine condition in the Paris markets in early spring. In the first instance three or five little plants may be put under each glass, and these thinned out and used as they grow, so that eventually but one is left, and that, without exaggeration, often grows nearly as big as the glass itself. Happily, no water is required, as the ground possesses sufficient moisture in winter and spring, and evaporation is prevented by the glasses and the protecting litter that covers the space between them. Thus a genial, agreeable moisture is kept up at all times, and the very conditions that suit lettuces are preserved by the simplest means.

By means of the same glasses the various small saladings may be grown to perfection, or receive a desirable start. Thus, for instance, if corn salad be desired perfectly clean and fresh in mid-winter, it will be obtained by sowing it between the smaller lettuces grown under these glasses; and so with any other small salad or seedling that may be gathered or removed without loss before or at the time the more important crop requires all the room. These bell glasses will be found of quite as much advantage in the British garden as they are in the French; they will render possible the production of as fine winter salads in our gardens as ever the French grew; they will enable us to supply our own markets with an important commodity, for which a good deal of money now goes out of the country; and, not least, their judicious use will make fresh and excellent salads possible in winter. At present the produce is so inferior and so dirty at that season, that it is generally avoided, and rightly so; for lettuces when hard and wiry from alternations of frost, sleet, and rains—slug-eaten and half-covered with the splashings of the ground, above which they hardly rise—are not worth eating or buying. And though they may be grown well in frames and pits, the method herein described is better and simpler than that, and the lettuces thus produced are finer than those grown in English gardens in winter.

The Barbe de Capucin is the most common of all salads in Paris in the winter and early spring, and for its culture the *cloche* is not required. It may be readily raised by every person having a rough hotbed, or even a cellar, and is a good salad that nobody need be without. It is perhaps too bitter for some tastes, but is often now used by English families, and is well worthy of culture in small gardens, because so very easily forced when other salads are scarce. The market gardeners in the neighbourhood of Paris sow in the month of April every year quantities of the wild chicory seeds on purpose to produce this salad. Before the early frosts they dig them out with a fork, taking great care not to break the roots, and put them in by the bells in a place where they will be ready to hand in the middle of October. A hotbed is prepared in an obscure place—in a deep cellar or cave without air or light will be preferable. In this genial hotbed they are forced: being first tied up in bundles, slightly sprinkled with water, the tops cut off at the end of a fortnight or three weeks, and the old roots thrown away, as they will not bear a second cutting. They may be kept for planting out, but it is scarcely worth while to do so. Successions of this are kept up every fortnight.

This salad is of all others that which may be had with the least amount of trouble by any person in possession of a spot of rough ground, a cellar, or any dark place where a little heat might be used to start the blanched leaves of the chicory in winter; and therefore it is desirable that it be brought into universal use. Should the taste be too bitter to those unaccustomed to it, or who do not like bitter salads, the addition of corn salad, celery, beetroot, &c., improves and modifies the flavour, and makes a very distinct and agreeable salad. Space prevents me going more into the various particulars as to salad culture in the neighbourhood of Paris, where a great number of varieties suited to the various seasons are grown, and there is in consequence some slight variations as to treatment, but the important and chief points of their cultivation are indicated in the preceding pages. It should, however, be added that endive, lettuce, &c., are cultivated and forced to a considerable extent, and with great success, on the narrow hotbeds elsewhere alluded to.

The general state of the Parisian market gardens is such that I cannot conclude without making some allusion to it. It has been frequently said that the minute division of property in land retards the improvement of agriculture in France. It may be so with farming, but it certainly does not hold good with market gardens. Those in and around Paris are comparatively small, but they are the best and most thoroughly cultivated patches of ground I have ever seen. Every span of the earth is at work; and cleanliness, rapid rotation, deep culture, abundant food and water to the crops—in a word, every virtue of good cultivation—are there to be seen. I doubt very much if such good results could be obtained by a larger system, and certainly in no part of Britain is the ground, whether garden or farm, so thoroughly cultivated or rendered nearly as productive as in these little family gardens. They may be so called, for they are usually no larger than admits of the owner's eye seeing the condition of every crop in the garden at once, and the French market gardeners as a class " keep to themselves," marry among themselves, and seem content with about as much ground as gives occupation to the family. They are as a rule a very prosperous class. I am not aware that their superior culture has been noticed by any former horticultural writer who has visited France; but Paris being within eleven hours of London, the truth of what I say of its market gardens can be readily tested, and they are at all times worthy of being seen.

MUSHROOM CULTURE.—In old times the market gardeners of Paris used to grow the mushroom with profit, but since the *champignonnistes* cultivate it with much less danger from cold in the caves under Paris and its environs, the market gardeners, who used to raise it to a great extent in the open air, do so now to a much smaller degree. In fact, they would have to renounce its culture altogether were it not that the mushrooms they produce are whiter, finer, and bring a better price than those grown in the caves. They begin with the preparation of the manure of course, and collect that of the horse for a month or six weeks before they make the beds; this they prepare in some firm spot of the market garden, and take from it all rubbish, particles of wood, and miscellaneous matters; for, say they, the spawn is not fond of these bodies. After sorting it thus, they place it in beds two feet thick, or a little more, pressing it with the fork. When this is done the mass or bed is well stamped, then thoroughly watered, and finally again pressed down by stamping. It is then left in that state for eight or ten days, by which time it has begun to ferment. After these eight or ten days the bed ought to be turned well over and re-made on the same place, care being taken to place the manure that was near the sides of the first-made bed towards the centre in the turning and re-making; then they leave the mass for another ten days or so, at the end of which time the manure is about in proper condition for making the beds that are to bear the mushrooms. If they do not find the stuff "sweet," unctuous, and of a bluish-white colour in the interior, they do not expect much success; but by carrying out the foregoing simple directions there is little chance of having it otherwise. Then they make the little ridge-shaped beds—about twenty-six inches wide and the same in height—formed like "the back of an ass," and placed in parallel lines, at a distance of twenty inches one from the other. The manure is made into close little beds gradually and carefully, the man pressing it down well with the fork, so as to give the whole mass a firm close-fitting texture, so to speak, and gradually narrowing as he builds till his little ridges are finished. Of course the length of those ridges will be determined by the wants of the grower; in a market garden they may extend over and cover a considerable extent. The beds once made, the manure soon begins to warm again but does not become unwholesomely hot for the spread of the blanc or spawn. When the beds have been made some days, the cultivator spawns them, having of course ascertained beforehand that the heat is genial and suitable. Generally the spawn is inserted, the holes being made in one line around the bed, within a few inches of the base, and at about thirteen inches apart in the line. Some cultivators insert two lines, the second about seven inches above the first. In doing so, it would of course be well to make the holes for the spawn in an alternate manner. The spawn is inserted in bits about the size of three fingers, and then the manure is closed in over, and pressed firmly around it. This done, the beds are covered with about six inches of clean litter. Ten or twelve days afterwards they visit the beds, to see if the spawn has taken well. When they see the white filaments spreading in the bed they know that the spawn has taken, and that it is good. If they do not see that it has begun to spread, they do not leave the bed alone, as too many do amongst ourselves, but take the spawn they suppose to be bad and replace it with better. But, using good spawn, and being practised hands at the work, they rarely fail in this particular; and when the spawn is seen spreading well through the bed, then, and not before, they cover the beds with fresh sweet soil to the depth of about an inch or so. When the beds are made in the open air—as, indeed, they generally are—the little pathway between them is simply loosened up, and its soil applied equally, firmly, and smoothly with a shovel. With these open-air beds they succeed in getting mushrooms in winter. The *chemise,* or covering of hay or litter, is put on immediately after the beds are earthed, and kept there as a protection. They have not long to wait till the beds are in full bearing, and when they are in that state it is thought better to examine and gather from the beds every second day, or even every day where there are many beds. Into each little hole from which mushrooms have been pulled they place firmly a small quantity of earth. Sometimes, when the *chemise* gets mouldy and rotten during a wet season, they remove it, and replace it with another; and sometimes during a very dry season they are obliged to water it. And thus they grow excellent mushrooms, and in great quantity.

The culture in a cave differs but little from the preceding. The accompanying cut shows the size and position of the little beds in the cave. Shelves may be raised against the walls to accommodate small beds like the narrow one placed against the wall of the cave. The manure is first prepared above ground, and the beds are covered with soft light earth—

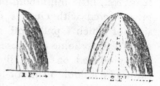

FIG. 73.

soft white chalky earth covering such as I have seen. So far as I had opportunities of observing, mushrooms are produced on those diminutive beds as freely as upon those of a larger size— as abundantly as could be desired, in fact; but I am not able to point out any distinct advantage upon the methods practised by good cultivators among us, and chiefly speak of it in consequence of the interest attaching to the cave-culture. I had, however, not sufficient opportunity of seeing the French system of mushroom culture. It is not very easy to visit the best market-gardens, and still more difficult to find one's way into the mushroom-caves.

Mr. James Barnes, of Bicton, who was as well acquainted with the old school of horticulturists as he is well known among the new, writes as follows of what he saw of mushroom-growing in the neighbourhood of Paris:—" Early on the first morning of my late visit to Paris, myself and a companion started to see the grea. central market. Our first object was to inspect the flower, fruit, salad, and vegetable departments, where we saw quantities of freshly-gathered mushrooms of fine quality. We then took a hasty run through the fish, flesh, fowl, frog, and snail departments, in which were fine fat snails huddled together by the bushel. Many an English cultivator would, I am sure, be glad to see his store of these pests pressed into such close quarters. Soon afterwards a cab conveyed us to the entrance-gate of the horticultural department of the great International Exhibition. Here we feasted our eyes on fruits, vegetables, salads, &c., and also made a close inspection of the fruit-tree department, where many and various methods of training, &c., were displayed. Here were to be seen placed about the vegetable-tents, &c., little mushroom-beds in the form of Lilliputian potato-pits, covered with clusters of mushrooms of good quality, at the sight of which I pulled off my hat and began to rub the dust out of my eyes. Some one close by asks, Have these mushrooms really been produced on those little toy-looking beds? Oh, yes, there cannot be a doubt of it, for here are buttons bursting up, varying in size from that of a pea to that of a child's halfpenny ball ; and on inspecting them afterwards I found these bunches, which I had first seen as buttons, were open plates, the size of a teacup, some even larger than that. Of course, the character of the materials of which the beds consisted did not escape my observation. I also saw, one evening afterwards, near the Palais Royal, in the window of a splendid restaurant, fine pineapples, melons, pears, grapes, and other fruits, as well as salads, &c., exhibited, and along with them one of these miniature mushroom-beds—aye, and covered over with good mushrooms, too. 'Well,' muttered I, 'what an old goose I am! after cultivating mushrooms for fifty years, not to have thought of this.' When a boy I used to be sent to a farmer's dunghill in search of mushroom spawn ; and we brought home at times a cartload or so of natural spawn in the shape of lumps, spits, or sods, from mill track, hotbed, or elsewhere, and often it was in a damp state. To secure it for future use, it was placed in sheds to dry slowly ; but on that next the wall, or in damp corners, lots of mushrooms would burst forth, and exhaust the spawn, if not occasionally moved about. Some of these lumps and crumbs of spawn placed in boxes, and cased over, set in gentle heat, did, in fact, produce many good dishes of mushrooms. Also under the footpath trellises of early vineries and peach-houses I have seen lots of natural spawn placed and covered, from which mushrooms were gathered at the commencement of forcing, when the heat was only moderate. Notwithstanding, however, these hints, received from nature herself, blindness continued, and no idea whatever occurred to me that this useful esculent could be so easily produced for the million as I now find it can be. Let me now state the fact that the little French mushroom-beds consisted of flakes or sods of fine natural spawn cut out of the stable dung-heap, encased in a light-coloured, loamy, muddy soil—not, however, battered, rammed, or trodden hard, as is usual with us, but laid over the surface to about the same thickness as that of an ordinary mushroom-bed, and left a little loose and knobbly—an advantageous condition as far as watering is concerned.

"As regards mushroom-growing, my mind was, however, not yet exactly at ease, so I kept rummaging about day after day in my gardening tours through royal and other extensive forcing grounds. Where there was a good deal of stable-dung, I always noticed the smell of mushroom spawn, and the dung placed in tidy-shaped heaps full of it. This seemed to me to be very generally the case, and that from these heaps any quantity of fine, strong, natural spawn could be cut out in flakes or sods of any size or thickness, for laying on the floors of caves, cellars, or sheds, or on mushroom-house shelves. Mushrooms could thus be cultivated in any cave or cellar, old box, tea chest, or the like.

"How is it that this beautiful, strong, natural mushroom spawn is so abundant in France? Can it be because entire horses are the rule there, not the exception, as with us? Let us suppose a quantity of natural mushroom spawn to be discovered in a dunghill, hotbed, mill track, &c. Atmospheric influence prevents this body of spawn from being utilized where it is found. The custom, therefore, has been, on discovering such useful treasure, to collect as much of it as retains its freshness and full natural properties, in flakes, sods, or cakes, and to gradually dry it for spawning newly-made beds. It must be gradually dried, but not in the full sun, or in a quick draught; for, if so, it will surely perish. Instead of submitting it to this drying process, however, were we to place it at once on the floor of a cave, cellar, mushroom-house, or in any other suitable situation, quantities of mushrooms would be the result in a few days. It is by simply watching and assisting nature in her interesting ways that such useful results occur. For instance, immense quantities of natural spawn are allowed to perish and exhaust themselves for lack of observation or the simple knowledge of how to utilize it. Twenty tons may be hidden in a dunghill, and through the unsuitableness of the atmosphere not a hatful of mushrooms may ever be produced therefrom; but had this spawn been discovered, and placed in situations in which a suitable atmosphere could have been commanded, sufficient mushrooms might have been produced to feast the inhabitants of a large town, and that at a very reasonable rate. Even supposing we were in a locality in which natural spawn could not be readily discovered, a suitable dung or muck heap could be spawned, and when in good condition turned to valuable account, as previously described. From what has been said it will be clear that we are yet very blind to nature's ways, and that we have, moreover, much to learn.

"I am aware that it has been considered difficult to produce fine crops of mushrooms of the best quality throughout the year, on account of our sudden atmospheric changes; for example, during the past season few natural mushrooms have been produced. Mushrooms will not stand either heat, cold, wind, wet, or drought. A dry summer, succeeded by fine early autumn rains, not excessive; these followed by quiet, warm days and humid, foggy nights, when the warmth of the atmosphere by day is from 60° to 65°, and then only for a short time at noon, and when that of the nights ranges from 50° to 55°—these are the conditions under which the greatest abundance of good mushrooms is sure to be produced naturally. Taking, therefore, the year through, the cave, cellar, and

mushroom house proper are the only places in which these conditions can be fully borne out; but for many weeks during spring,
summer, and autumn mushrooms might be freely produced in
various ways, and in many places, very simply, indeed, with
but little trouble or expense, by those who have at command a
cellar, a shed, the corner of a fuel house, a tent, or some other sheltered corner."

CHAPTER XVIII.

Horticultural Implements, Appliances, &c.

APART altogether from modes of culture, we may effect considerable improvement, neatness, and economy in our gardens by adopting several articles described in this chapter, and I would earnestly draw the attention of both professional and amateur gardeners to their claims.

THE CLOCHE.—This, which is simply a large and cheap bell-glass, is used in every French garden that I have seen, and it is the

FIG. 74.

cloche which enables the French market gardeners to excel all others in the production of winter and spring salads. Acres of them may be seen in the market gardens around Paris, and private gardens have them in proportion to their extent—from the small garden of the amateur with a few dozen or scores, to the large garden where they require several hundreds or thousands of them. They are about sixteen inches high, and the same in diameter of base, and cost about a franc a piece, or a penny or two less if bought in quantity. As the "moonist" of poor Artemus Ward failed with his moon, so my artist has failed with the *cloche*, but in another direction—he has made it too ornamental, given it an elegant rim, and rendered it altogether too good-looking. The form in common use is almost straight sided and somewhat conical. In reply to numerous correspondents who have written inquiring about this cheap bell-glass, I have to state that the English dealers have not got them, nor are they likely to procure them, getting as they do several times the price for the bell-glasses in common use with us for indoor propagation, &c. But, nevertheless, I trust that the *cloches* will not be kept in the background for that, as they are re-markably cheap, and as useful for indoor or any other propagation as the very costly articles now in vogue with us—apart from their use in salad growing, &c. Messrs. Vilmorin, the well-known seed merchants of Paris, have obliged me with the address of a person who supplies them—Rouchonnat jeune, 75, Rue du Fau-bourg St. Antoine. He offers them at 85 francs per 100, if more than 500 are taken; smaller quantities at 90 francs per 100—*i.e.* at the rate of about ninepence each. The *cloches* are packed by twenties, four francs being charged for the package; but the vendor will not be responsible for breakage in transit. The advan-tages of the *cloche* are—it never requires any repairs; it is easy of carriage when carefully packed (one inside the other in a rough frame made for the purpose); by carefully handling, one is very rarely broken in the Paris market gardens—level as a billiard-table and without a leaf out of its place; they are easily cleaned—a swill in a tank and a wipe of a wad of hay every autumn clears and prepares them for their winter work. They are useful in many ways besides salad growing; for example, in advancing various crops in spring, raising seedlings and striking cuttings: and finally, they are cheap. A thousand of them may be bought for 40*l.*, or less in France, and with good management these would soon more than repay the cul-tivator. But of course it is only in market gardens that they would be required in such quantities as that, and in some small gardens not more than a few dozen will be wanted. Every garden should be furnished with them according to its size; and when we get used to them and learn how very useful they are for many things, from the full developing of a Christmas rose to the forwarding of herbs, and even stools of asparagus in spring, I have no doubt they will be much in demand. Having spoken of the most particular and important use of this article in connexion with salad culture, no more need be said of it here.

EDGINGS FOR PARKS, PUBLIC GARDENS, DRIVES, &c.—The edgings in gardens have a very important bearing on their general aspect, and often on their cleanliness. Hosts of people with gardens are continually looking out for a good edging, and many are taken in by the aspect of those made of tiles, &c. Any variety of brick, imitation stone, or terra-cotta edging, is the ugliest and most unsatisfactory thing that can be admitted into an ornamental garden. Massive edgings of stone around panels, &c., in geometrical gardens are of course not included in the pottery ones now alluded to. These last are enough to spoil the prettiest garden ever made, and are as much at home round a country seat as a red Indian at a mild evening party. Look at them as they are carefully arranged by the exhibitors in one or two of our public gardens, and possibly you may think they are clean, symmetrical, and the very thing desired. When brought home and arranged around the borders their true charms begin to display themselves. Being "of a geometrical turn," if I may be allowed so to describe them, they must be arranged so as to look quite straight in the line. If they wabble about, one this way and one that, they "don't look nice," even granting that the things themselves are tolerable. Now it is nearly impossible to "set" them easily and cheaply, so that they will remain erect. To have them set by a mason may be resorted to by some; but it is simply a way of wasting money. Of course, a good workman may set them neatly enough by ramming down the soil firm on each side; but, even if perfectly well arranged, they are, after all, the worst variety of edging known. Then, again, they are often of a texture that cracks into small pieces with the first frost, though there are some much more tenacious. The expense in the first instance is a good deal, and one way or another they become unsatisfactory, till there is no tolerating them any longer, and they are thrown by with the old iron or the oyster shells.

The reason why people have resorted to them is, that the edgings ordinarily used prove disappointing and dirty, and they long for something that will be neat and tidy at all times. To abuse a bad thing without offering a better, or any at all, is often better than to stand still and tolerate a nuisance; but in this instance I am able to recommend a capital permanent edging—everlasting, in fact, and with nothing that could offend the most critical taste. This is simply made of rustic rods of cast iron, in

FIG. 75.

imitation of the little edgings of bent branches that everybody must have seen. They are evidently cast from the model of a bent branchlet, generally about as thick as the thumb, but they are of various sizes. The marks where the twigs are supposed to have been cut off are visible, and altogether the thing looks as rustic as could be desired, is firm as a rock when placed in position, and, in a word, perfect. These irons are of course stuck in the ground firmly, and as shown in the figure. But, while prettier than any stick edging ever seen, they are, when fastened, also the most firm and permanent of all. They may be set up by any boy. The fact that they are not stiff and ugly tile-like bodies prevents their offending the eye if one or two should fall a little out of place here and there. But this is nearly impossible; for at the place where every two sticks cross each other they are tied by a little bit of common wire. They should be so plunged in the walk, or by the side of the walk, that about seven inches of the little fence appears above ground. This, however, may be varied with the size of the subjects which they are used to encompass; six or seven inches is the height given for edges for ordinary purposes. They are equally useful for the park, pleasure-ground, or even the kitchen-garden. In parks and pleasure-grounds, however, we usually have edgings of grass, and therefore it may occur to the reader that they are useless therein; but the little fences of bent wood which furnished the idea for these iron edgings were generally used to prevent grass near much-frequented spots from being trodden upon; and of course those now recommended will answer the purpose better. But it is in much-frequented places along drives, and in public gardens and parks, that their chief merit will be found. They may be seen in every public garden in Paris, from the little squares near the Louvre where you may notice them obscurely running along outside of the ivy edgings, to the slopes of the Buttes Chaumont and the more frequented parts of the Bois de Boulogne, and they must ere long be as widely adopted in England, for it is impossible to find a better or more presentable edging.

WATERING IN PARISIAN PARKS, GARDENS, &c.—The French system of watering gardens, &c., is excellent, or at least the generally adopted system; for at the Jardin des Plantes there are yet watering-pots made of thick copper, which are worthy of the days of Tubal Cain, but a disgrace to any more recent manufacturer, and a curse to the poor men who have to water with them. Generally Parisian lawns and gardens are watered every evening with the hose, and most effectively. It is so perfectly and thoroughly done, that they move trees in the middle of summer with impunity; keep the grass in the driest and dustiest parts of Paris as green as an emerald, the softest and thirstiest of bedding plants in the healthiest state; and as for the roads, the way they are watered cannot be surpassed. They are kept agreeably moist without being muddy, while firm and crisp as could be desired. Of course all this is effected in the first instance by having the water laid on; but that is not all. With us, even where we have the water laid on, we too often spend an immense amount of labour in distributing it. In Paris generally it is applied with a solitary hose, which pours a vigorous stream, divided and made coarse or fine either by turning a cock, by the finger, or even by the force of the water. This is of course the way they apply it to beds of shrubs, &c., that require individual watering, so to speak, and it is also the way they water the smaller bits of grass about the Louvre and other places; but when it comes to watering a large space of grass it is very different, and then it is that their system is seen in all its excellence. One day in passing by the racecourse at Longchamps I saw it carried out in perfection. The space had become very much cut up by the great review and the races. In any case they water it to keep it as green as possible in summer. At first sight it would appear a diffi- cult thing to water a racecourse, but two men were employed in doing it effectually. Right across the whole open space from east to west stretched an enormous hose of metal, but in joints of say about six feet each. The whole was rendered flexible by these

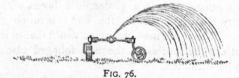

FIG. 76.

portions being joined to each other by short strong bits of leathern hose, each metal joint or pipe being supported upon two pairs of little wheels. Fig. 76 shows a section of the apparatus at work. By means of these the whole may be readily moved about without the slightest injury to the hose in any part. At about a yard or so apart along this pipe jets of water came forth all in one direction, and spread out so as to fully sprinkle the ground on one side; and thus four feet or so of the breadth of the whole plain of Longchamp was being watered from one hose. There were two of these hoses at work, one man attending to each of them; the only attention required being to pass from one end of the line to the other, and push forward the hose as each portion became sufficiently watered. In this way also I saw them watering the lawns at Baron James Rothschild's beautiful place in the same neighbourhood. The simplest thing of all is the way they make the perforations for the jets along the pipe. They are simply little longitudinal holes driven in the pipe with a bit of steel. They must be made *across* the pipe, or the water will not spread in the desired direction. There are various modes of spreading water in use about Paris, but none of them half as good as this simple method. The hose for watering the roads is arranged on wheels also, but, as it must be at all times under command when carriages pass by, it has only one rose or jet, which is directed by a man who moves about among the carriages with the greatest ease, and keeps his portion of the road in capital condition. Of course it is a much cheaper way than carrying the water about as we do, as then we must have horse and cart, wear and tear, and man also; whereas, by having the water laid on, all the men have to do in watering is to attach the hose and commence immediately. In the same way as much work can be done in a garden in a day as with ten men by the ordinary mode; so that in the end it is much cheaper to have the water laid on. There can be no doubt that to the efficient watering much of the success of the fine foliaged plants in Paris gardens is to be attributed.

The profuse cropping of the market gardens here is also due, to some extent, to their regular watering: the ground being always moist and friable. In the market gardens they raise the water from deep wells by horse-power, and send it to barrels and cement cisterns plunged in various parts of the garden. In each market garden, and in many nurseries, you usually see a horse of the establishment working one of these pumps. The remarkably succulent radishes, turnips, &c., with which the Paris markets are supplied in the heat of summer, owe their merit to abundant watering and a very rich light soil. Indeed, they sow them on beds nearly composed of well rotted manure.

ATTACHING WIRE TO GARDEN-WALLS, &c.—If there be any one practice of French horticulturists more worthy than another of special recommendation to the English fruit-grower, it is their improved way of placing wires over walls or in any position in which it may be desired to neatly train fruit trees. Than our own mode of erecting espaliers, or wiring walls, nothing

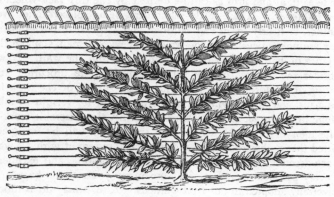

FIG. 77.

can be more hideous. So many have been the failures in British gardens as regards the placing on walls of the wire to which to affix the trees, that it has been given up as useless and too expensive, and many have said that the old-fashioned shred and nail are the best things. But there is a very much better and sounder way, and I am completely converted as to the value of the French mode of wiring a wall, of which Fig. 77 is an illustration. In the first instance, several strong iron spikes are driven into the wall at the ends—in the right angle formed by two walls, and then rough nails, or rather hooks, are driven into the wall in straight lines by a mason,

S

exactly in the line of direction in which the wire is wanted to pass. The wires are placed at about ten inches apart on the walls, and the little hooks for their support, also galvanized, are placed at about ten feet apart along each wire. The wires are made as straight as a needle and as tight as a drum, by being strained with the raidisseur. Through the hooks the wire is to pass for its support, and of course it is necessary to have those lines of hooks quite straight, a thing which a mason can readily arrange. The wire —about as thick as strong twine—is passed through the little hooks, fastened at both ends of the wall, into the strong iron nails, and then made as tight and firm as can be desired by the tiny rack-wheel of the tightener. These little tighteners being placed one over another, or nearly so, in vertical lines, and at great distances apart, they are not noticeable on the wall. The wires remain at about the distance of half an inch or three quarters from the wall. If we consider the expense of the shreds and nails, the procuring and cutting of the former, the destroying of the surface of the walls by the nails, and the leaving of numerous holes for vermin to take refuge in, the great annual labour of nailing, the miserable work it is for men in our cold winters and springs,—it will be freely admitted that a change is wanted badly. The system of wiring a wall above described is simple, cheap, almost everlasting, and excellent in every particular; and it must ere many years elapse be nearly universally adopted in our fruit gardens. A man may do as much work in one day along a wall wired thus as he could in six with the old nail and shred. As to galvanized wire having an injurious effect on the fruit trees trained on, it is simply nonsense; I will not therefore waste my space and the intelligent reader's time by discussing it. That is the only thing that can be said against this beautiful system of wiring. Given a concrete wall as described elsewhere in this book, smoothly plastered and wired thus, what fruit trees could be in a more excellent position than those upon it? The temporary coping taken off after all danger from frost was past, every leaf would be under the refreshing influence of the summer rains, all the advantages of walls as regards heat would be obtained, the syringing engine would not be counteracted by count-less dens offering dry beds and comfortable breeding-places to the enemies of the gardener and the fruit tree, while the appearance of the wall would be all that could be desired.   We will next turn to an equally important subject—

TRELLISES FOR ESPALIER TREES.—The French call their wall-trees " espalier," and in adopting the word from them the English have transferred it to those standing away from walls, but trained in a manner similar to wall trees.   They term our espalier *contre-espalier*, but the terms wall-tree and espalier are distinctly and generally understood among us, and therefore it is better to use them in the usual sense.   I merely mention this to account for using the word espalier in its English sense while discussing matters connected with French fruit culture.   The simplicity and excellence of their mode of making supports for espaliers will be better shown by Figure 37 and others than by verbal description.   The raidisseur is again employed here—one for each length of wire, and the strong end supports are of T-iron.   One whole garden is devoted to single and double trellising of this kind at Versailles—the tall lines of trellising being placed at about five yards apart, and rods of strong wire run from one to the other, so as to give the whole square of trellises mutual support.   As neat dividing lines in English gardens, and as the best support ever introduced for espaliers, such trellises are sure to spread amongst us.

The galvanized wire which is so universally useful for the fruit garden is sold in twenty-three different sizes.   Of this an inter-mediate size is that best suited and usually selected for strong and permanent garden work, albeit a mere thread to the costly bolt-like irons we use.   The sort suited for walls is sold at about 3*l*. 6*s*. for 100 kilogrammes equal to a little more than 220 lbs. English.   Each kilogramme (a trifle more than 2 lb. 3 oz.) of this affords more than 131 English feet of wire.   The price given is that for the second quality of wire, the first quality of the same pattern costs about 6*s*. 6*d*. more for the 220 lbs.   Thus, of this wire of the very best quality, and such as if placed properly in its position is as permanent as it is useful, 200 lbs. avoirdupois may be obtained for less than 4*l.*, and 220 lbs. will extend a distance of 13,123 English feet. This size will also suit well for espaliers, though a size or two stronger had better be used for the tallest and strongest trellises. A size, or even several sizes smaller, will suffice for dwarf trellises with three rows of wire or so, to accommodate very dwarf trees of any shape the cultivator may desire.

Since writing the above, I have been informed by Mr. J. B. Brown, of 90, Cannon-street, and 148, Upper Thames-street, E.C., that he supplies the chief Parisian houses who furnish French fruit-growers with this useful galvanized wire, and therefore we may expect to procure it cheaper in England.

USEFUL TRELLISING FOR YOUNG TREES IN NURSERIES, &c. —In some of the best fruit nurseries I noticed a simple and effective kind of trellising used for training young wall and espalier trees. It is useful in enabling the French to keep in

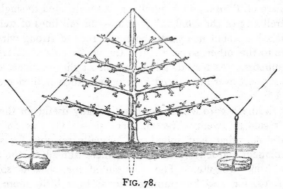

FIG. 78.

stock trees for these purposes to a greater age than is the case in our own nurseries, and for various purposes should prove useful to the grower of young fruit trees. I need not say that a larger and modified application of the same plan would do well

for large espalier trees, indeed I have seen it applied with good
effect, and it perfectly suits a method which is not uncommon in
France, of keeping the upper branches of trees, trained horizontally,
shorter than the lower ones, so as to secure perfect vigour in the
lower branches. This trellis may be established at a trifling cost
by using light posts of rough wood, or, if permanent and greater
strength be desired, of T-iron. In either case the posts must be
firmly fixed. The wire should be passed through a hole or strong
eye in the top of the pole, and fixed with stones or irons in the
ground. For the better training of the shoots straight, their rods
may be extended from the post to the wires with but little trouble.

TRELLISING FOR THE PEACH TRAINED AS AN OBLIQUE CORDON
AGAINST WALLS.—The wire and the little straightener is very
efficiently used so as to do away with any necessity for nailing in

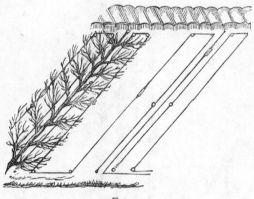

FIG. 79.

training the peach *en cordon*, as shown in the accompanying figure.
When the lines which the wires are to follow are fixed upon bolts
and eyes are driven in, the wire is fixed to and passed through them,
and then tightened with our useful little friend the raidisseur.

PROTECTION FOR ESPALIER TREES.—This is a thing we rarely attempt, but which the French sometimes do with success.   The

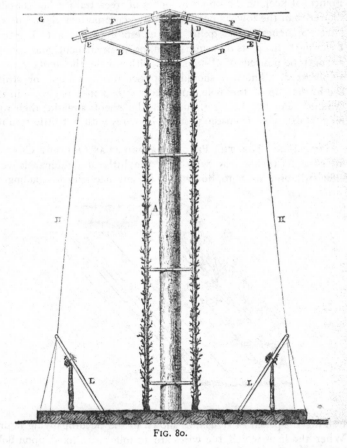

FIG. 80.

accompanying figure shows a mode of arranging two rows of trees in a manner different to that already shown on the double trellis.

Here the main supports are strong posts. French gardens are usually surrounded by walls, and in establishing a system of trellising for growing the choicer pears it is considered wise to stretch an occasional wire from the trellis to the adjoining walls or from one trellis to another. Thus if a whole square is devoted to galvanized trellises for pears—at say nine feet high, and at from fifteen to twenty feet apart, the intermediate space being cropped, the trellises in addition to being individually well supported, afford each other a mutual support by means of strongish wires running *across* all the lines of trellising, say at thirty feet apart. This is shown by the line G, in Fig. 80. At the bottom run rows of horizontal cordon apples of the most important kinds. The posts are placed closer apart in erecting the trellises than when the trees are "abandoned" to the vicissitudes of the weather.

B B shows the support for the temporary protection, F F; galvanized wires are run through about the points E and D, while the lines H H, descending at long intervals, are fixed firmly to stones in the ground and to the little iron posts, L L. Neat straw mats are generally fixed on the top of these, the mats being made so neatly and firmly that no untidiness is observed. For the English garden, however, tarpaulin on cheap light frames would be better. I describe this more for what it suggests than anything else. Some like arrangement is badly wanted with us, and need not be difficult to contrive. By having a few lines of choice apples trained on the low-cordon system at each side, and two good rows of pear trees, a great deal of valuable fruit could be protected at the same time by making some arrangement whereby the whole could be covered with cheap canvas.

COPING FOR WALLS.—Having several times spoken of the deep copings the careful French cultivator uses for his fruit wall, I here give a rough figure showing a section of the tile-coped wall, and projecting from beneath the coping the iron rod which serves as a support for the temporary coping. The artist has omitted to show the rod slightly turned up at the end. The French take a good deal of trouble with temporary copings, and find them of the greatest use in getting good and regular crops, for the frosts are severe in the northern parts and all around Paris, in fact nearly all the region north

of the river Loire, the most important region of France. The best temporary coping I have ever seen used was narrow lengths of tarpaulin nailed on cheap frames from six feet to eight feet long, and about eighteen inches wide. The use of such on walls devoted to the culture of choice pears, peaches, &c., would result in a marked improvement. The temporary coping has a great advantage

FIG. 81.

in being removable, so that the trees may get the full benefit of the summer rains when all danger is past, and not suffer from want of light near the top of the wall, as they would if such a wide protecting coping were permanent. I believe that such a coping would be much more effective than any of the netting and canvas protections now in use in English gardens.

MATS FOR COVERING PITS, FRAMES, &c.—In our cold and variable climate, the winter covering for many minor glass structures is of the greatest importance. It is a thing that at present we do in a very expensive and by no means satisfactory way. The French mode of doing it is much cheaper, neater, and more effective; and in passing through their market gardens and forcing-grounds in winter, it is one of the first if not the chief thing that seems to the English horticulturist as worthy of imitation. The covering used consists of neat straw mats about an inch thick, the sides as clean and neat as if cut in a machine, the mat knit together by twine, and its texture such that it may be rolled up neatly and closely. If I mistake not, one of these mats, which is much better as a protecting agent than a bass-mat, costs about one-third the present price of a bass-mat, while in point of appearance and amount of protection given the advantage is all in favour of the French *paillaisson*. The figure given represents a simple frame for making these mats in the nurseries of M. Jamain, the celebrated cultivator of orange trees, and I append his description of it. There are several frames for this purpose, and there is also a machine for making these mats, which are indispensable to the French gardener; but the one here described is the best and simplest for general use. " Get two pieces of timber (1) about 3 inches thick, 4 inches wide, and as long as required. Pierce these timbers, as shown in the figure, and introduce A in the holes to maintain the same width between the sides, and support the nails or screw, as shown in the cut. These nails are to keep the string tight (5). The board may be shifted from hole to hole so as to make mats of any desired length. The length of the string must be about three times as long as the straw mat, and rolled round a little handle, made as in E. The straw must be placed on the machine so as to have all its cut or lower ends close against the sides, the tops meeting in the middle, and so thick as not to have the straw mat

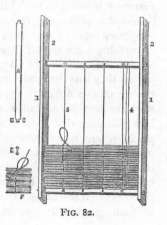

FIG. 82.

thicker than three-quarters of an inch when finished. The stitches must not be wider than three-quarters of an inch, and be worked as follows (see F of the figure). Take a little of the straw with the left hand, and work the handle with the right, first over the straw, then over the bended string, coming back underneath, and passing swiftly the handle between the two strings, pulling tightly and pressing the straw, so as to have a flat stitch, and not thicker than three-quarters of an inch at the most. The same operation is repeated until the mat is finished. The machine described has been at work for the last twenty years in our nursery, at Paris, and is yet as good as new. An ordinary workman may make daily from thirty to forty yards run of these straw mats with it.'' Wherever frames and pits have to be covered, the adoption of these mats would effect a considerable saving, and be productive of much satisfaction to the cultivator.

All new or strange things of this sort are adopted slowly by horticulturists; but that they would immediately adopt this, if they had an opportunity of seeing it in working order, I have no doubt; and I hope yet to see it in general use in British gardens. In France it is found so useful that it is employed for many purposes besides that of covering frames, and they even make a very effective temporary coping for walls of it in some cases. I doubt very much if anything I can say for these mats will give the gardener a full idea of their utility. In all gardens where men are regularly employed they may be made during bad weather in winter; and as there is often a difficulty about procuring enough of useful indoor work for men during bad weather, the making of these mats will be a gain from that point of view alone. Indeed, in many country places, where straw is abundant, the making of these mats will cost a mere trifle. As bass-mats are now very expensive compared to what they were, this matter is well worth the attention of horticulturists generally. The mat is shown in use at Fig. 83, a little thicker than it should be. In connexion with this subject, the system of framing generally used in France may be alluded to.

FRAMES FOR FORCING, &c.—The French market gardeners use an immense quantity of frames, and it is by their aid they procure most of the tender and excellent forced vegetables, &c., sent to the markets in early spring. These frames are made of very rough wood; are narrow—not exceeding four feet in width; and arranged in close lines completely immersed in the heating material. The illustration will give an idea of the appearance of these beds.

Most undoubtedly the principle is better and cheaper than our own. We employ large and well-made frames in private gardens, and for the most part place them so that all but the base is exposed to the influence of the weather, and the plants therein are more liable to changes of temperature and cold. By having the frames narrow, all the sidework rough and cheap, and the frames placed in close lines, we get the greatest amount of heat at the smallest cost. By having nothing but the surface of the glass exposed, nothing is lost, and when the frames are covered by the neat, warm, and flexible straw mats above described, they are as snug as could be desired. When it is simply desired to preserve bedding plants, &c., through the winter in frames, the spaces between the rough-sided frames are merely filled up with leaves and slightly heating material, as shown in my figure; but where used for forcing, the manure

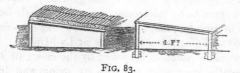

FIG. 83.

is of course placed underneath the frames also. About two feet of space is left between each frame, or just enough for the convenience of the workmen. Generally they seemed to me to be put together by the workmen of the market gardens : two stout posts being driven firmly in at one end, and an end-board nailed to them. Then at every four feet or so minor posts are driven down, and the rough front and back-boards nailed to them. The cropping of these frames is dense and excellent, as is the case with the market garden cropping generally.

GLAZING GLASSHOUSES.—In the matter of glasshouses we have little or nothing to learn from the French beyond what is displayed at La Muette. A great variety of houses were shown at the Exposition, all of iron, usually neatly and well built, occasionally elegant, but offering very little which is worthy of adoption. The

French use iron houses, and generally think our wooden ones too heavy and cumbrous. One mode of glazing is worth notice, and better and neater than we often see in England. The panes did not overlap, but met evenly, a particle or so of putty being used between them if they do not meet exactly ; and on the outer side a neat thin strip of lead paper, about half an inch wide, is laid over each junction. The appearance of this silvery strip is good, and it is in every way effective. It is, in fact, heartily adopted by M. Barillet, and is probably not surpassed, if equalled, by any mode of glazing. The house is by its use glazed almost hermetically, while there is not a particle of the rustiness that occurs in English houses from decaying putty, &c., to be seen. There is no difficulty whatever in the repairing of such houses, which is not the case in some of our novel modes of glazing. This method is good and elegant : the best for large houses, which should generally be of iron, if we do not wish them to be intolerably ugly. The small houses of the French are of iron too, and nothing can be more useful than the very cheap, two-light, low iron houses at La Muette. The aspect of most French glasshouses is spoiled by the effect of the shading employed—laths connected together by string and little hooks, and painted dark green, and by nearly all the houses above the small and useful low two-light ones having a gangway over the roof, so that men may pass to arrange this shading, &c. It is a bad, very awkward, and very expensive way. The mode of glazing here alluded to is well worthy of our adoption for conservatories, and indeed all iron houses, and may be seen largely employed in the great nursery garden of the city of Paris at Passy.

GARDEN LABELS.—There were very many labels shown in the gardens of the Exposition, the best being some small ones exposed by M. H. Aubert, 189, Rue du Temple. These were in rather thickish zinc, and very neatly and deeply impressed and accentuated. They were of sizes suitable for rose and fruit trees, small ones for numbering, others with stems for using with pot plants, and so on. There is a thinner type of the same style common enough in our seed-shops, and not very dear: but after you get them comes the writing, and that is the difficulty, or the stamping, or whatever it may be—usually a failure, and always a labour requiring some taste and neatness. Now you can get these labels deeply, and neatly, and plainly impressed by this M. Aubert at three francs per 100, furnishing him with a list of the sorts required in the first instance. As anybody who has meddled with such matters knows, to get such labels deeply and well impressed would be very cheap at three or even five francs the 100, even without considering the cost of the labels themselves. I need therefore hardly say that being able to procure them ready to attach to roses, fruit trees, &c., at three francs per 100 is a great advantage. Our labelling generally is a great mistake, and this surely is a step towards having plants decently labelled without incurring great expense. Small ones for numbering are sold at one franc per 100.

GARDEN CHAIRS.—We will next turn to chairs, of which there are numerous kinds to be seen. The kind of chair which may be seen in quantities in all public places in Paris, with a convex seat made of elastic strips of metal springing from the sides and joined together in a little central piece, was much admired by English visitors. Fig. 84 represents it. There are many modifications of this, the best being one in which the seat made of these united bands is covered with a slight wire network. These chairs stand any weather, and are nevertheless as elastic and comfortable as any drawing-room chair. The neatest and most elegant and comfortable conservatory, pleasure-ground, or summer-house chair ever seen is composed of three of these seats united in one, the larger framework of the back and sides being made of charmingly rustic iron about

FIG. 84.

as thick as the thumb, the smaller spray being tied to the larger by imitation osier twigs. This was shown by M. Carré, the maker of the greater number of chairs in this way.

TRUCKS FOR TUBS OR VERY LARGE POTS.—This very handy little truck is what the French use for moving orange tubs, &c. It would be impossible to find anything more useful in its way.   I

FIG. 85.

have seen the gardeners actually running with big specimens on them. The pot or tub is caught by the little iron feet, then thrown on its side and tied firmly if a long distance has to be traversed.

THE NUMEROTEUR.—Numbering instead of labelling is now adopted in so many gardens and nurseries, that the making widely known of this useful instrument cannot fail to be useful. The following description of it originally appeared in the *Gardeners' Chronicle* :—" Horticulture is a science so vast, and embraces subjects so different, that however good a man's memory may be it is insufficient, and hence it becomes necessary to give it mechanical aid. Among the means employed are tickets or labels written upon parchment or paper, or small pieces of wood, &c. ; but these are soon effaced, and are very liable to get lost or displaced.   A very good plan frequently adopted consists in the use of small bands of lead, which are rolled round the stems or branches of the plants. Upon this lead a number is marked, corresponding with a catalogue, in which the name and any particular remarks are entered.   This method is sure; but to carry it out several things are necessary. First there is wanted a series of numbers from 1 to 10, or rather from 1 to 9, the zero combined with other figures, making the numbers 10, 20, 100, &c.   Then this series of numbers must be fixed upon a block of wood, and the figures have to be impressed

upon the leads by means of a small hammer. So that to mark the
leads we want—1st, a pair of scissors to cut the metal; 2nd, a set

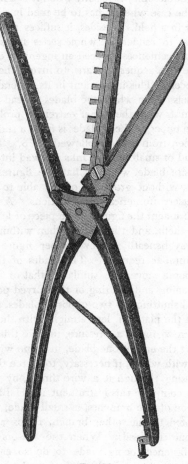

FIG. 86.

of numbers; 3rd, a block to receive them; and 4th, a hammer to strike and indent the figures in the leads.   This apparatus therefore becomes troublesome, especially when it is necessary to change its place, as is the case when it has to be used in different parts of a large garden, or in a field.   Besides, it suffices for one of the little figures to be lost to render the whole series useless.   A consideration of these inconveniences induced an ingenious cutler, M. Hardiville, of the Rue St. Jacques, in Paris, to invent the Numéroteur, or Numbering Pincers.   This instrument in its general form resembles a large pair of scissors, in which the blades instead of being cutting are flat and blunt, with the upper extremity prolonged.   On the inner side of the upper of these blades is fixed a series of ten figures arranged in order, from 1 to 9, followed by 0.   These figures are placed at the end of small steel shanks screwed into the blade, and upon the opposite blade, which is flat, the figures are marked in hollows, so that, without grouping, one is able to effect with certainty any necessary numerical combinations.   A pressure of the blades suffices to indent the figure in the piece of lead that has been placed between them, and the lead is then withdrawn and placed in the same way beneath whatever other figure or figures may make up the number required.   The blades of these numbering pincers work upon a movement similar to that of a pair of scissors, the alternate opening and shutting of the curved portion or handle also opening and shutting the two opposite blades, so that it is only necessary to put the plate of lead straight with the figure which is wanted, and then to make a pressure, to have this figure indented on the lead.   At the end of one blade, in a line with the figures, is a small punch, with which, if necessary, to pierce the lead, in order to admit of passing through it a wire thread, by which it may be suspended.   To complete this instrument M. Hardiville has added, on the side of one of the branches, a small blade, which by means of a spring adapted to the other branch, forms a pair of scissors with which to cut the leads.   When the scissors are not needed, the spring is unfastened, being made to do so easily and quickly, and the blade then tightens itself against the branch of the pincer without any trouble. At the base is a moveable spring which serves to open the branches. Thus we see that this instrument is very complete, but its value is augmented by its not being complicated, and especially by its being of a reasonable price— ten francs."

THE SÉCATEUR.—Of garden cutlery I will only mention the *sécateur*, and this is an instrument that every gardener should possess himself of at once. I know well the prejudice that exists in England among horticulturists against things of this kind, and their almost superstitious regard for a good knife—a state of things quite justified by the nature of most inventions brought out for the amelioration of gardening. I also believed in a good knife above all, but when I saw how useful is the secateur to the fruit-growers of France, and how much more easily and effectively they cut with them exactly as desired, I became at once converted. A sécateur is seen in the hands of every French fruit-grower, and by its means he cuts as clean as the best knife-man with the best knife ever whetted. They cut stakes with them almost as fast as one could count them; they have recently made some large ones for cutting stronger plants—such as the strong awkward roots of the briars collected by the rose-growers. Of these sécateurs there are many forms, several of the best being figured here. First we have the Sécateur Vauthier (Fig. 87), a strong and handy instrument. Its sloping semi-cylindrical handles have their outer side rough, which gives a firm hold; the springs, though strong, resist the action of the hand gently; the curvature of the blade and the adjustment are perfect; and lastly, the principal thing, the action is so easy as never to hurt the hand.

"During the many years of my experience," observes M. Lachaume, a fruit-grower who describes this implement in the *Revue Horticole*, "I have used tools of all kinds, and the tools have also used me a little; but I have never met with anything which gave me so much satisfaction as the Sécateur Vauthier. Every desirable quality is combined in it, and I recommend it with perfect confidence. The

T

strongest branch will not resist its cutting, nor a single branch, however well concealed, be inaccessible to it.   Moreover, the double notch on the back of the blade and hook (in which a wire

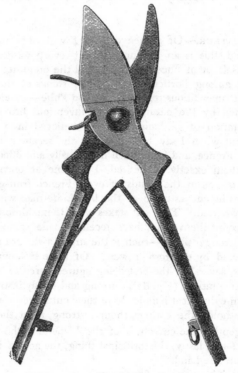

FIG. 87.

is shown in the figure) will enable the operator when employed at his trellises to cut every wire without using the pincers.

The Sécateur Lecointe (Fig. 88) is another variety recommended by the leading French horticultural journal.   It appears to have some advantages over the others in the substitution of a coiled spring for the

ordinary flat spring previously adopted. The inventor was led to
devise this kind of spring in order to avoid the annoyance arising
from the frequent breakage of the form
usually employed. It is said that this form
of spring secures an easy and gentle action
of the instrument, and has the advantage
of lasting longer than others, from not
being so liable to break, while it secures
a firmness and evenness in working which
is not otherwise attained. A further improvement is pointed out in the fastening,
which consists of a stop which catches
when the two handles are drawn together,
a projecting portion on the outside acting
as a spring which is to be pressed when
the instrument is required to be opened.
M. Lecointe of Laigle is the inventor.

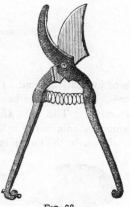

FIG. 88.

Fig. 89 represents the sécateur of older
date than the preceding, and one more
generally used. It is much employed at Montreuil. There can
be no doubt that where much pruning of any kind is done, and
particularly pruning of a rather rough nature,
the sécateur is a valuable implement. It was
first invented by M. Bertrand of Molleville.

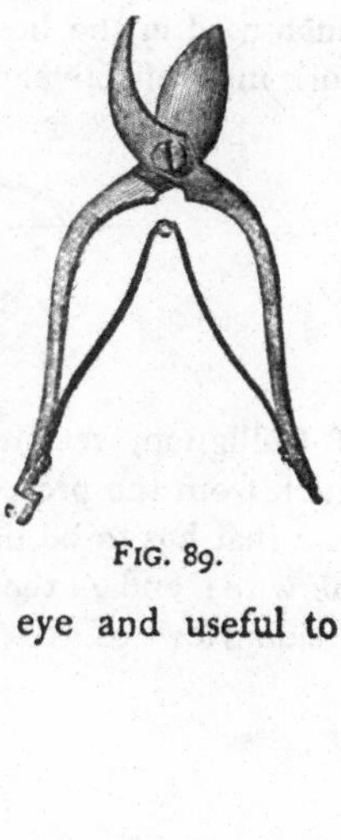

THE RAIDISSEUR.—This is the name for
the little tightening, straightening machine,
which plays such a very important part in the
wiring of garden walls, or erecting of trellises
for fruit-growing in France. It is an implement which, though insignificant in itself, is
calculated to make a vast improvement in our
gardens and on our walls. It will save labour,
time, expense, and make walls, permanent
trellises, &c., infinitely more agreeable to the eye and useful to

FIG. 89.

the cultivator than ever they were before. There are various forms
which I need hardly describe, as they are so well shown in the accompanying cuts. The first (Fig. 90) is a reduced figure of one about

FIG. 90.

three inches long, and of which I brought some specimens from
Paris. The engraver has placed it in the best position to show its
structure. The wire that passes in through one end is slipped
through a hole in the axle; the other end is attached to the tongue,
as shown in the engraving, and then by the aid of a key, Fig. 91,

FIG. 91. .

placed on the square end of the axle, the whole is wound as tight
and straight as could be desired. The first form figured is very
much used in the best gardens, and always seemed to me to do its
work most effectively. The next figure is the *tendeur*, or stretcher

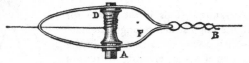

FIG. 92.

of Collignon, recommended by Du Breuil. It does not differ
much from the preceding. D shows the point of insertion of the
wire that has to be tightened; B the fastening of the other end of
the wire; and A the head on which the key is placed. Fig. 93 is
a side view of the same implement. The best form of these

raidisseurs cost about threepence apiece in Paris. We cannot well have more efficient things, though we may succeed in getting them of a cheaper form. Indeed, I once saw one in use, about **forty**

FIG. 93.

miles from Paris, which could not have cost more than a good nail, but unfortuately I forgot to secure a specimen or take a sketch of it. However, I hope to secure it some day. The foregoing kinds are galvanized, just like the wire. That shown by Fig. 94 is a **very**

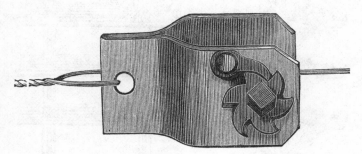

FIG. 94.

simple one, not galvanized, and which was shown in use in the fruit garden of the Great Exposition. This last form is surely such as can be readily and cheaply produced in any manufacturing town. The best of these tighteners are insignificant in price; and if it were not so it would still be profitable to employ them in consequence of the great saving they effect, by enabling us to use a very thin wire, which is quite as efficient and infinitely neater than the ponderous ones now generally used with us, where the nail and shred has given way to some costly system of wiring.

IMPROVED FRUIT SHELVES.—The stocks of winter pears I had
the pleasure of seeing in some French gardens were remarkably fine
in many instances; but, in one case, the mode of storing them in
the fruit room was so superior that I have devoted a couple of
figures to illustrating it.   Instead of being confined to wide shelves
or benches all round, as is usually the case, there were several sets
of shelves arranged along the room—rather narrow, sloping oak
shelves—supported by oak uprights.   These shelves are wide
enough for five rows of pears on each side, and on such a slope that

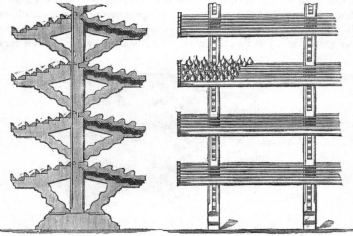

FIG. 95.

the pears rise gradually, line after line, so that the eye could see
each fruit with ease, without handling or disturbing any, and of
course this was a great gain.   But the careful constructor had gone
further, by making the slight concavity upon which each line of
pears rested of two laths, so that the air could flow up beneath them.
No single fruit was allowed to touch its fellow, and thus they were
in a very much better condition than in the British fruit-room,
where all are frequently packed tight together, and the good ones

often liable to get tainted by the bad. This was in the pear-room at Baron Rothschild's, and a more pleasing sight could not be presented to the lover of a garden—the successive shelves of splendid fruit being so arranged that every individual pear could be examined without touching one. I need scarcely say that in the case of a fruit requiring so much nicety of judgment and attention as the pear does, in the gardener who makes the most of his collection, and has each kind, or even each perfect fruit, eaten at the right time, this is a great gain.

A NEW LAWN PLANT.

AMONG the more important plants which I became acquainted with in the Jardin des Plantes for the first time is one whose name it is not expedient to mention oftener than is absolutely necessary—Pyrethrum Tchihatchewii (Boiss.), and which I forgot to allude to while describing that establishment. I first saw it growing on the mound near the large cedar, covering the dry and arid ground under the shade of trees with a close carpet of short, fresh, dark green leaves, elegantly divided, and lying quite close to the ground. The position was a sloping one, the soil chalky and very bad. Notwithstanding these drawbacks, nothing could have looked better than the little carpet formed by the Pyrethrum. It not only looks well in summer, when grasses, if they cannot grow well in such positions, are at all events to be seen in a fresh green state in others; but in winter and early spring, when all around is dead and at rest, it looks as green as healthy ivy. It is perfectly hardy, and runs about close to the ground in a remarkably rapid manner. It does not seem to grow more than an inch and a half or two inches high, and in spring and summer bears white flowers, which rise a little higher than those of the daisy, and must be removed in the same way if the making of a fine lawn with this plant be attempted. From what I have seen of the plant, I have no hesitation in saying that it is certain to be useful in some positions in gardens and pleasure-grounds, and probably will be found generally useful. The spots for which we may count upon it as a gain are deeply-inclined banks on a dry soil, where grasses grow with difficulty, and perhaps almost perish during dry weather in summer; the many places under trees, which are usually bare from the shade, and from the roots of the trees drying and impoverishing the surface. From the numerous inquiries that are made about suitable plants for covering such positions, there can be no doubt that this will prove a great boon to many for these purposes alone; and on dry sandy hot soil, where grass perishes in summer, it may prove the plant so long sought for on such. The Spergula, so famous at one time, and which was so great a failure generally, from its superficial hold of the surface and tendency to die off in patches, and become spotty after attaining a green and luxuriant condition, has no merits compared with the Pyrethrum, which "takes hold of the ground" in a determined way, and from its fine constitution is not at any time liable to go off from disease. Having no experience of its doings on a rich loam or a clay soil, I cannot say what it may prove worth on such; but it is worthy of a trial on all. On slopes, among other positions in which a loose and shifting soil was troublesome, it is very likely to prove valuable, the well divided leaves matting amongst the sand, which serves to hold it well together. When distributed amongst cultivators, trials of its use, both alone and in company with other turf-making plants, should be made. It is not as yet to be had for sale in this country; but I hope some of our nurserymen will introduce it ere long.

FRUIT TREE PROTECTION, CORDONS, &c.

DURING the latter part of April, and after this book had gone to press, I made an extensive tour through the midland and northern counties, visiting many of the best-managed gardens, taking particular notice of the state of the garden fruit crops, and more particularly of the means taken to protect them from frost.

By this time many fruit-growers must be aware of the disastrous effects of the spring upon their crops of wall fruit. We have rarely had a spring so destructive. At first bright and beautiful for many weeks, it induced the trees to waken out of their wintry sleep before the accustomed time; and then, just as pears were fully in bloom, and apple trees blushing with the unfolded flowers, killing frosts occurred and destroyed the expanded blooms of the pear, the young closed buds of the apple, and to a greater extent those of other fruits. This was the case about London, and also to a considerable extent in other districts. Never had bloom buds been so abundant. Some may suppose their apples to be safe, while the flower buds are in numerous cases killed at the heart, though perfect looking externally. Such at all events is the case in low-lying situations, those more or less elevated seeming to escape with slighter injury.

If I had nothing to say of the subject but this, it would simply be a waste of space to notice it; for the loss of fruit crops in our gardens is too common an occurrence to demand recording. What does require notice, and the urgent attention of everybody interested in the garden culture of fruit trees, is the fact that the rule in British gardens is not to afford the trees any efficient protection while they are in a state liable to be injured by frost. I have recently visited many of the finest and best-managed gardens in England, and found the best and tenderest of our open-air fruits without protection in the majority of cases, and very imperfectly protected in others. Gardens grand in extent and with every

reasonable means, display their walls without any noticeable coping
to throw off cold rains and sleet, and protect the trees from frost.
In many cases the very imperfect protection of some kind of netting is afforded to the peach trees (which do not require it so badly
as other kinds), and all the rest are " left to nature" with a vengeance.
In the case of ordinary standard or orchard trees, things must
be left to take their chance; but it is little less than disgraceful
that this should be the rule with trees for which we go to considerable expense to build walls and pay constant attention to. By planting orchards with crops beneath the trees, as the London market-
gardeners do, we insure a crop in any case; and paying but
little or no attention to the trees when failure does occur, the loss
is not great.   Not so with the wall and garden trees.   Their culture is little better than an expensive delusion until we take care to
protect them, so that a crop will be secured every year.   The present system is not only deplorable from the direct loss of fruit that
it occasions, but also from the fact that the trees, having nothing
to bear during a season when the blossoms are destroyed, make
wood so gross and infertile, that, even if the following spring prove
a favourable one, there may be a poor crop or none at all from want
of fruit buds. Walls are expensive; wall trees, if properly managed,
are expensive too, for they, unlike an orchard tree, require skilled
attention, and a great deal of it; and they can only pay by taking
care to guard the fruit buds from death by frost.   There is no finer
climate than ours for the perfecting of wall fruit; the difficulty
presented is to get the fruit well set and out of danger of frost. We
have good seasons and bad seasons—years in which the blossom
opens early and is killed, years in which it opens late and is killed
also, and others in which some kinds of fruit escape without injury;
but the right way to avert loss of crops is to give up all guessing
about good or bad seasons, realize that we are always liable to such
a destructive spring as the present, and make such arrangements as
will render ourselves quite fearless of the like.

The destruction is apparent in the best-managed gardens as well
as the worst. The following extracts from the report of the able

superintendent of the Chiswick Garden, Mr. A. F. Barron, published in the *Gardener's Chronicle* of April 18, will explain how matters stand in the Chiswick Garden:—" I have made a pretty general examination to-day (April 14) of our fruit prospects, and have to report as follows :

" *Pears.*—Bloom very abundant on all varieties. Flowers large and healthy, but fully two-thirds destroyed by frost on the night of March 24, when the thermometer fell to 19 degrees. Some of the earlier varieties on dwarf pyramids and espaliers were then in full flower, such as Louise Bonne, Doyenne d'Eté, &c. Of these scarcely a blossom escaped; others which were then only in bud did not suffer so severely. This present week is again exceedingly trying for the blossoms, the latter being much excited by intensely warm sunshine during the day, with the nightly occurrence of from seven to eight degrees of frost, and biting easterly winds. Under these circumstances I find a little more injury done every night. The relative effects of frost on different varieties, under exactly the same conditions, are worth notice.

" *Apples.*—Bloom abundant; considerably injured when in bud from the same cause as that which affected pears, especially in the case of dwarf pyramidal trees. It may be worthy of remark that apple blossom is much more tender than that of the pear. Apples in the bud state are just as easily destroyed as full-blown pear blossom, and both are more tender than peach blossom.

" *Plums.*—Blossoms abundant; nearly all destroyed on dwarf trees, and slightly injured on tall standards.

" *Cherries.*—Bloom abundant; greatly injured, especially on dwarf trees.

" *Peaches* promise well : in short they are not at all injured even where they have been unprotected.

" *Apricots* very scarce—destroyed by frost.

" It is sad, however, to find our expectations of a bountiful yield of fruit thus destroyed, while one short month ago they were so bright.''

It will thus be seen that no garden can be in a worse plight for want of protection than that of the Royal Horticultural Society, which ought to be an example to the country. The wall pears and other fruits in the gardens of the Royal Horticultural Society, and in the majority of gardens, instead of being protected from the frost, are perhaps more exposed to its influence than the flowers of a standard tree. The tree allowed to assume its natural shape has its blossoms arranged in so many ways—some low in the tree and protected by upper branches, some well exposed to the sun, and others on the late and shady side—that it has a good chance of setting a sufficiency of fruit. The pear grown against a wall, and equally exposed to the sun in all its parts, is, on the other hand, induced to expose its blossoms more at one time; and the spurs projecting beyond the usually insignificant coping of the wall against which the tree is placed for heat and protection, it is a mere chance if the blossoms escape destruction. So it is with other fruits. "Peaches promise well" at Chiswick and elsewhere—doubtless because the close way in which the slender shoots of the peach are nailed to the wall secures it some protection from sleet and frosts, no matter how insignificant the coping.

Compared to this system of half-doing, that of the London market-gardeners growing standard trees and cropping all the ground underneath with bush fruits and vegetables, so as to secure a crop in any case, is excellent; and if they generally took the trouble to thin the branchlets, so as to obtain larger and more perfect fruit, would leave nothing to desire, so far as kinds that grow well as standards are concerned. Of the present state of wall-fruit culture in this country, and of the necessity for improvement, there can be no doubt. The remedies are simple and certain. The adoption of the wide temporary coping of tarpaulin, nailed over light frames about eighteen inches wide, described elsewhere in this book, and which was the best and simplest and neatest of the various copings I saw in use in France, would alone result in a vast improvement to our crops of wall fruit. It is so light, so easily placed, and so effective. Fixed close under whatever permanent coping the wall may possess, and sloping down with a roof-like pitch, all cold, rains, and sleet are thrown effectually off; and it prevents radiation more effectively than the best protections now in use, even in gardens where the walls are considered worth the trouble of protection in spring. It may be fixed so as to be undisturbed by the highest winds; and when the trees are out of flower may be taken away, stored without trouble for another year, and the walls then exposed to the refreshing and cleansing summer rains. No permanent coping of like breadth and efficiency is advisable; it would darken the wall too much, whereas when the temporary coping is removed every leaf enjoys full light, and the wall may be perfectly covered from top to bottom with healthy wood.

But to fully compensate us for the trouble of thoroughly protecting walls, a little revolution is necessary in our garden fruit culture. We must, as elsewhere pointed out in this book, crop the borders in front of walls with fruit trees trained on the low and simple cordon principle. By doing so we shall at once dispose of the much-debated question of what is best to do with the fruit borders, and at the same time collect such a valuable lot of fruit trees immediately in front of each wall, as would render it both convenient and highly desirable to protect efficiently both walls and borders, and by the same means. A very narrow border would accommodate four or six lines of apples or pears trained on the cordon principle, and an extension of the cheap canvas protection given to the wall would suffice to cover them also. Where the borders were wide a greater number could, of course, be planted in parallel lines, and some simple mechanical arrangement made, whereby both wall and border might be covered.

In the report from Chiswick Mr. Barron remarks, that "a spring like this would be fatal to low cordons." This, coming from such a quarter, seems to invite a word of reply from me. Considering that nearly everything else was killed, it was fair to assume that the "low cordons" would be likely to share the lamentable fate of the general stock, even though there is not one in the Chiswick garden. I have frequently seen low cordons, properly managed, bear abundantly in districts of France as much exposed to spring frosts as the neighbourhood of London. Low cordons *might* lose their crop such a season as the present, but what variety of fruit tree can be so easily protected if arranged as I have elsewhere in this book described—the walls and borders both perfectly covered and the whole thoroughly protected? By proper management the severest spring we experience could not injure such cordons; and the only thing that their extensive planting would prove fatal to would be reports such as those that are now too common, and of which the above is a type. Even Chiswick, much as it suffers during such frosts as the present, could not fail to succeed with the cordons arranged as proposed. I will venture respectfully to suggest to the Council that by planting its fruit borders with cordons, and thoroughly protecting all, they would by the certain and superior crops soon obtain a considerable return, and at the same time teach the public a useful practical lesson.

Whether the border is covered with cordons or not—rising but little above the ground, they cannot shade the wall—the lower parts of the walls, if not perfectly covered with the larger trees, should be covered with cordons as soon as possible, and from them alone may be gathered, in many places where there is bare wall space, as much fine fruit as would pay for the expense of protection. The only objection urged against my proposal, that the bare lower portions of our garden walls, &c., should be covered with the finer apples and pears, worked on the Paradise and Quince stocks, was that if the walls were properly and sufficiently clothed with the trees to which they are devoted, there would be no room for the cordons. A likely objection enough, if walls were often to be seen in such a state. I have recently noticed *miles of garden wall quite bare at the bottom which is capable of affording quantities of as fine fruit as it is possible to grow.* And if I noticed this in a tour through the richest and best gardens in England, how much more so does it apply to the thousands of gardens in these islands that do not receive regular and skilled attention! However, my proposal for rendering this waste space highly profitable is certain to be universally accepted ere long; and I was agreeably surprised to find it already under trial on a large scale in the gardens at Welbeck, the magnificent seat of the Duke of Portland.   Mr. Tillery, the able gardener, has every confidence in its success, and his little Calville Blanche apples are in full bloom, though not planted many weeks.   Thus my hope that we shall soon furnish our own supply of this fine kind will probably be fulfilled sooner than expected.   Nothing can be more easily established or more pleasant to attend to than these little trees of choice apples and pears on the bottoms of low walls, walls of houses, pits, &c., with a sunny exposure.   And though the idea of parallel lines of neatly furnished cordons stretching along the fruit borders may not seem practical to all, the day is not far distant when its real advantages will be clear to every cultivator of our choicer hardy fruits.

INDEX.

LONDON :
SAVILL, EDWARDS AND CO., PRINTERS, CHANDOS STREET,
COVENT GARDEN.

Printed in the United States
By Bookmasters